Advance praise for

FRENEMY **NATIONS**

"Soderstrom takes us on a global journey to neighboring countries that seem like twins separated at birth, looking back through their family albums to understand how they grew into what they are today. Along the way, the reader is provided with something that all books should aspire to: a better understanding of the world we live in."

—Wade Shepard, author of *On the New Silk Road: Journeying through China's Artery of Power*

"Original, wide-ranging, and insightful, *Frenemy Nations* is a towering work of literary nonfiction. Here we see a highly skilled writer in action, using first-person narrative and voice to control a centrifugal whirl of scholarship."

—Ken McGoogan, author of *Flight of the Highlanders: The Making of Canada*

"Mary Soderstrom's 'frenemy nations' are lessons in how to love one's neighbor, how geographically close regions can generate mutual benefit, and above all, how positive change is possible."

—Rita Shelton Deverell, television broadcaster and author of *American Refugees: Turning to Canada for Freedom*

"An insightful and fascinating narrative of borderlands that examines many of the world's most distinctive national neighbours and friends, *Frenemy Nations* demonstrates that even where common borders and geographical proximity create shared experiences, paths and choices differ. Soderstrom's attention to the importance of context and place makes this a fresh and compelling read, but also a cautionary tale about national relationships and the continuing importance of place."

—Heather N. Nicol, author of *The Fence and the Bridge: Geopolitics and Identity along the Canada-US Border*

LOVE AND HATE BETWEEN
NEIGHBO(U)RING STATES

FRENEMY
NATIONS

MARY SODERSTROM

University of Regina Press

Printed and bound in Canada at Friesens. The text of this book is printed on 100%
post-consumer recycled paper with earth-friendly vegetable-based inks.

Cover design: Duncan Campbell, University of Regina Press
Text design: John van der Woude, JVDW Designs
Copy editor: Ryan Perks
Proofreader: Donna Grant
Indexer: Sergey Lobachev, Brookfield Indexing Services

Cover art: "Map Location Icon" by 1001/iStock Photo, and "Map of Standard Time
    Zones of the World " (altered) by Calvin Dexter/ iStock Photo.

Library and Archives Canada Cataloguing in Publication

Title: Frenemy nations : love and hate between neighbo(u)ring states / Mary
    Soderstrom.
Names: Soderstrom, Mary, 1942- author.
Description: Includes bibliographical references and index.
Identifiers: Canadiana (print) 20190113987 | Canadiana (ebook) 20190114002
    | ISBN 9780889776722 (softcover) | ISBN 9780889776876 (hardcover) | ISBN
    9780889776739 (PDF) | ISBN 9780889776746 (HTML)
Subjects: LCSH: International relations. | LCSH: Geopolitics. | LCSH: Boundaries. |
    LCSH: Borderlands.
Classification: LCC JZ1329.5 .S63 2019 | DDC 327—dc23

10 9 8 7 6 5 4 3 2 1

University of Regina Press, University of Regina
Regina, Saskatchewan, Canada, S4S 0A2
tel: (306) 585-4758 fax: (306) 585-4699
U OF R PRESS   web: www.uofrpress.ca

We acknowledge the support of the Canada Council for the Arts for our publishing
program. We acknowledge the financial support of the Government of Canada. /
Nous reconnaissons l'appui financier du gouvernement du Canada. This publication
was made possible with support from Creative Saskatchewan's Book Publishing
Production Grant Program.

For Lee, of course...

CONTENTS

THAT TRIP
TO MONTREAL

I n the summer of 1968, in the middle of the American involvement in Vietnam, my husband and I loaded up our vw Beetle and immigrated to Canada so he could accept a job offer in Montreal. We were young, we were disgusted with the war, and we were hopeful that we'd find something different across the border.

But, to be honest, we didn't expect things to be too different. After all, weren't Canada and the United States very much alike? They were both supposedly freedom-loving democracies with a high standard of living and a history of free speech. The differences we knew about all sounded like pluses. Lots of people spoke French in Montreal where we were headed, but that would only make things that much more interesting. Beer had a higher alcohol content: this, too, was all to the good. And then there was the big, important difference: Canada was not involved in Vietnam.

Of course it was not as simple as that. In later years I've often said that I had six months of not worrying about what was happening in the States, only to end up, as I learned more about my new home, with two countries to fret about and to try to change.[1]

Which led to a long-running reflection on what makes these two countries different, and by extension, what makes other places that have so much in common they might seem like unidentical twins, continue as separate entities. Many of these places I've visited, others I've worked hard to try to understand, all I've come to care about greatly.

Just south of Montreal lie two examples: the states of New Hampshire—home of "Live Free or Die" populism—and Vermont—home of Bernie Sanders and his peculiarly American brand of socialism.

Farther west in Canada I soon discovered two other examples: supposedly "redneck" Alberta and Saskatchewan, the cradle of Canada's universal healthcare system.

The stalwart vw. *Photo provided by author.*

Over the years, I came in contact with others:

Tunisia and Algeria, nations on the southern shores of the Mediterranean, whose histories give a glimpse of how colonialism has functioned for the last twenty-five hundred years.

Burundi and Rwanda, two countries in the Great Lakes region of Central Africa. Both have a major ethnic division, between the Hutu agriculturalists and the Tutsi herders. The two groups have at various times tried to exter- minate each other even though their language and religion are fundamentally the same.

Haiti and the Dominican Republic, which share the island of Hispaniola. Their different colonial experiences—Haiti was a French colony, the Dominican Republic, Spanish—have left their mark on both the land and the population.

Which brings to mind the great split in South America between the Spanish-speaking and Portuguese-speaking parts. That's a complicated story that sounds like something out of *Game of Thrones.*

And then there are the ancient island realms of Scotland and Ireland, whose histories are in many respects similar, and whose long shadows cover the globe. There are some who would say that the two countries were Britain's first colonies.

Two other neighbours that also saw British colonialism close up are the Indian states of Kerala and Tamil Nadu, who share the southern tip of that country.

Then there are the two countries of North and South Vietnam, no longer separated after the war that loomed behind our move to Montreal.

This book is an exploration of these pairs—these frenemy nations—and it culminates with a consideration of what makes the United States of America and Canada so different, as well as speculation about what brings on change in societies. Begun when Donald Trump was leading the pack in the run-up to the 2016 Republican nomination and just after Stephen Harper and his Conservatives were turfed out in favour of Justin Trudeau's Liberals, the book presents, I think, some reflections on change and culture that will be of interest to citizens of many countries.

Among the factors that make for differences are geographic variations, no matter how minimal; colonial history, which sometimes depends on a roll of the dice half a world away; exposure to the wider world; the traditional place accorded women; the shelter that a language can give; how the people are educated; and migration, both voluntary and involuntary. All of these have implications for the future of these unidentical twins, and for the fine lines, the borders, that divide them. They also carry lessons for those who'd like to change the future.

To begin, we'll take a look at the countries that started my reflection, the two Vietnams, and we'll end on a sunny August Sunday nearly fifty years later, with a celebration that marks, in a way, a happy ending, and which also underlines the way states and nations change.

THE TWO
VIETNAMS

These days it takes about half an hour to cross the mountain pass that divides northern and southern Vietnam. The road twists and turns, giving breathtaking views of the sea and mountains, and thrills and chills if you overcome caution and let your motorcycle go on the downhill side. Riding the pass on a motorbike has become one of Vietnam's must-do adventures. And even though the fellow travellers you encounter are likely to be only local delivery trucks and goats and cows, it's clear that the government authorities want to make sure that the experience is something to write (or vlog) home about. Red-and-orange triangle warning signs with an exclamation mark inside just emphasize the experience that is made marginally safer by the smooth, recently re-asphalted surface and the concrete blocks installed on curves to keep out-of-control vehicles from plunging over the side.[1]

At the top, besides spectacular views if you're lucky weather-wise (Hải Vân, the Vietnamese name, means "ocean cloud pass" because of the fog and mist that frequently obscure the coast), you'll find a fortress built in 1826 at a site that had served since the fourteenth century as the border between parts of what is now Vietnam.[2] In the late nineteenth century the French acquired it and rebuilt it, while in the twentieth century the Vietnamese themselves and then the Americans converted it to their own military ends. But today, on a sunny weekend, the parking lot at the top will be filled with tour buses and motorbikes and signs for Tiger beer, while brides in frothy white dresses sometimes pose with their sweeties on the circular roof of a bunker, looking for all the world like a bridal couple on top of a wedding cake.[3] More serious traffic rumbles underneath, through a tunnel that was opened in 2005, carrying goods and people from one side of a natural barrier to the other in a country that is now one of the most prosperous in Asia. A country unified since 1975, but one which has been divided more than once.

There were two Vietnams at war with each other back when my husband and I made our move from California to Montreal. Conflict had raged for more than twenty-five years when we left the United States in August of 1968. First there had been World War II and the Japanese invasion. Then came a lengthy war of independence against the French that ended in 1954 with the "temporary" division of the country into two zones, a north where the Communist Viet Minh held sway and in the south—the State of Vietnam—where a former

The curving road across Hải Vân Pass, a natural boundary between northern and southern Vietnam. *Source: Thibaut Schaller, Creative Commons Attribution 2.0 Generic licence.*

emperor governed. The dividing line was along the Bến Hải River to the north of Hải Vân Pass, which crosses a spur of mountains extending into the sea. On the north side cold "Chinese winds" blow in winter, while south of the pass, the weather is much warmer. Let poets make what they will of that divide (and they have), the line of division makes much more sense than the one that separated the Korean Peninsula at about the same time: in Korea, American generals actually took a *National Geographic* map and drew a line across it.[4]

Elections to unite the Vietnams were supposed to be held by 1956, but weren't because of opposition from the United States and the South Vietnamese.[5] There followed a civil war in which the North fought to take over the entire country under a Communist banner while the South battled for what was called the democratic option. The road that ran the length of the country—from Saigon, the southern capital, to Hanoi, the northern one— was the scene of battle after battle. With typical military gallows humour, American soldiers in 1968 were calling the portion that crosses Hải Vân Pass the Street Without Joy because army convoys crawling slowly up the steep, twisting road were sitting ducks for ambushes.[6] They'd borrowed the term from the French forces who had fought to control the highway fifteen years before. One of the major confrontations between French colonial forces and Vietnamese rebels, in fact, was called Operation Street Without Joy. It was designed to drive the rebels from a lowland stretch just north of the pass.[7] The operation wasn't successful, nor was the rest of the French effort. Nor would be the American war effort, as was becoming clear as we set out for Canada.

As it happens, there is film evidence of just how fraught the situation was. Television coverage of the fighting had made Vietnam the "living room war": correspondents on the ground reported daily for the big US news agencies. But there also exist "home movies." Shot by US soldiers, these short films were buried in archives for decades but have recently surfaced on the Internet. Eight of them—filmed, according to the clapperboard sequence that introduces each segment, by one SP/4 (Specialist 4th Class) Walker on the same days my husband and I passed through the American heartland—show a group of young men only a little younger than we were slogging their way north of Saigon. The first segment gives the view from a helicopter as it flies over flat, rice-paddy country on the Mekong Delta in the south of the country.[8] Then we see troops on the ground wandering through tall grass—or is it rice? The grunts wade through waist-deep swamp water, narrowly missing sharp bamboo stakes set as a booby trap. In a village, a group approaches a small house with louvred doors. One soldier, smoking a cigar, taps on a door, trying to open it. The men wait around until a woman hustles up and smilingly opens the door: all seems innocent.

But as the little films proceed, things become sombre. The grunts question a man about his papers, call in a helicopter to take someone out, and then, for reasons that are not clear, begin digging at the edge of a thicket. First they uncover a bone, like a small person's arm, then more bones, and then documents, and finally a skull still draped in wisps of skin. They spike it on a branch, take pictures of it. The scene shifts to soldiers wading across a stream, up to their shoulders, carrying their weapons above their heads. Waist deep in the big muddy and the big fool says to move on, as a song popular at the time put it.[9]

Unlike us, they were far from sure they'd get where they were going, wherever that was.

A US Marine moves a Viet Cong suspect to the rear during a search-and-clear operation in 1965. *Source: US Marine photo by Pfc/G. Durbin, public domain.*

— — —

At this point I must say that I've never been to Vietnam, but it has loomed large in my imagination for most of my life. In my writing, I try to report from the scene, as it were, but in this case, for various reasons, I haven't been able to travel there. Yet I want to include the story of the two formerly separate Vietnams in this reflection because, as we'll see, parts of the Vietnamese experience resonate with themes that we'll encounter again and again when we consider neighbouring places that are alike but different. That the two parts of Vietnam have been united for more than forty years now says a lot about how change can be effected too. And, as we'll see in the chapters about the United States and Canada, the consequences of conquest can have repercussions halfway around the world.

First, though, let's take a look at the map of present-day Vietnam. If you squint, the country's sinuous shape—the more prosaic might say it resembles an "S"—looks a lot like the profile of a dragon. The great head of the beast abuts China. It includes the flood plain of the Red River, one of the many life-giving rivers that rise thousands of kilometres away in the Himalayas, and which is as much a cradle of Vietnamese culture and civilization as the Yellow River is for the Chinese. On the map you could imagine the dragon's open mouth, ready to take a bite out of the Chinese province of Guangxi. Its plumed topknot encroaches into Thailand, while its tail runs down the eastern side of the Southeast Asian peninsula to its southernmost point in the Gulf of Thailand. Through the middle the mountains that form the peninsula's spine swing close to the South China Sea: the mountain spur running eastward is what the Hải Vân Pass crosses.

The countries into which the peninsula is divided today—Myanmar, Thailand, Laos, Cambodia, and part of Malaysia, in addition to the united Vietnam—are considered Mainland Southeast Asia. The region's northern boundary is open to discussion, but generally, the southern border of China delimits that frontier.

Back when we were leaving the United States, many thought that a victory by the Communist forces of North Vietnam would lead to all the countries of the region becoming Communist through the Domino Effect. This was during the Cold War, remember, a couple of decades after Mao Zedong led China to embrace Communism, and when the United States and the Soviet Union were each trying to win over the world, with China attempting to get into the game too. That conflict of ideas and powers skewed all political analysis. The truth is that the Vietnamese have had a healthy distrust of the Chinese for more than two millennia. Furthermore, when they were trying to set themselves up as an independent country after World War II, they had taken some inspiration from a document as American as apple pie, the Declaration of Independence, and, ironically, from the French Declaration of the Rights of Man and of the Citizen.

The Chinese controlled Vietnam from about 111 BCE to 938 CE,[10] except for a few short intervals when the Vietnamese temporarily shook themselves free.[11] One of the successful revolts was led between 40 and 43 CE by the Trung sisters, two women who still figure in the panoply of Vietnamese heroes. They ultimately were defeated, but it took thousands of Chinese troops to put down the revolt.

We know this early history in part because the Chinese assiduously documented their civilization. Unlike the writings of the Mesopotamian or Egyptian societies, which flowered at the same time as the first Chinese empire, ancient Chinese texts have been decipherable by the intelligentsia throughout China's history. While people in different parts of the greater Chinese sphere of influence may not have been able to understand each other when speaking, the educated have always been able to read Chinese texts. This is because Chinese writing is based on ideograms, each one representing a word that may or may not be pronounced the same in places a few hundred miles from each other.

And the folks who were living in the Red River Valley when the Chinese pushed southward spoke a language quite different from what is now Chinese. These people were descended from migrants from the islands of the western Pacific,[12] and Proto-Vietnamese is now classified as an Austroasiatic language, the language group spoken there. That fact may strike you as hopelessly arcane, even useless, but as we go along in this exploration of why places that are alike may be quite different, and why borders between them exist in the first place, language will come up again and again. What people speak says much, not only about where they come from, but also who they were colonized by and, occasionally, how they were able to stand up to forces from outside. But more about that later...

Today the Vietnamese language contains many loan words from Chinese, but its syntax is frequently distinctly *Việt* (as the Vietnamese would say).[13] One example of this is rather basic, something that anyone who has studied English and French can relate to: modifiers come before a noun in Chinese, as in English, but in Vietnamese they usually come after, as they do in French.[14] Nevertheless, the Chinese writing system was used for official documents in Vietnam and in the country's education system until the early twentieth century. A modified system, called *Nom*, used both Chinese characters and ones developed specifically for the Vietnamese context. (Following contact with Europeans, a Roman-style alphabet was developed, but more about that later on.)

At the time the Chinese and the Vietnamese cultures began to interact, the Vietnamese were already successful rice growers with a set of customs at odds with some important values held by the Chinese.[15] This meeting of two cultures is told in many stories, including one about two kings who disputed control of the Red River Valley. One, An Duong, was a migrant from China who had been established in the valley for some time. The other, Zhao To, was king of a province that included part of what is now southern China, and which was called Viet (or Yue, depending on your regional pronunciation.)[16]

Zhao, who had expansionist ambitions, had no intention of recognizing the other king, and so went to war. But armed conflict in those days wasn't a continual occupation, particularly in a region when troop movements and battles were nearly impossible during the rainy season. At one point a truce was declared and Zhao To and his son Shi Jiang visited An Duong's court. The young man fell head over heels in love with An Duong's daughter. His passion was returned and the young people married.

Up to this point, the story sounds familiar: a state marriage forges diplomatic ties, and everyone lives happily ever after (we'll see another example when Ferdinand marries Isabella and Spain is created). In this case, however, the story was more like Romeo and Juliet. Shi Jiang, even though he tarried in his father-in-law's court, was still a supporter of his father. His wife's father, he knew, relied on a magic crossbow to vanquish his enemies, so the young man went searching for it. When he found it, he destroyed it and then fled back to his father's forces. They attacked and subdued the king of the southern region, who was now bereft of his special weapon.

Defeated, An Duong fled toward the sea with his daughter in tow. But she still loved her husband, and—like Gretel in the European fairy tale—left behind clues along the way so he could find her. When her father realized what she was doing, he killed her and disappeared into the sea. It was sad news indeed for the young husband, and when he heard it, he killed himself so he could join his beloved in death.

Besides demonstrating that tragic love stories have been popular for a long time, this story tells two important things, according to commentators like K.W. Taylor. The first is that all the rulers of the land we now call Vietnam could trace their lineage to the Chinese, which gave them legitimacy during the years the Chinese had direct control over Vietnam. The second, although somewhat contradictory, is that the original Vietnamese society was different in a fundamental way from the Chinese one because it was matrilocal, not patriarchal. The Confucian norm was that the father's line was most important, and Chinese writers purported to be shocked that Shi Jiang had gone to live with his wife's family. In a well-ordered society wives lived with their husbands' families, where the pecking order started with the oldest male and then proceeded downward, with a son's wife near the bottom.

The custom was far from being just a sexist weapon. It also was the method by which the government and state were financed, since in this patriarchal system the family was responsible for paying taxes. But for a long time the folks living in the Red River Valley believed that riches should be passed through the maternal line, and that the whole community was responsible for paying tribute. During the millennium that rulers from China controlled the Viets they tried to stamp out the custom, but it died hard. After four hundred years of Chinese influence, one writer, referring particularly to family organization, noted that all the Chinese civilizing activity had had little effect: the Vietnamese "are on the same level as bugs," he wrote.[17] This may be the first time that an invading power tried to effect a *mission civilisatrice* in Vietnam, as the French would later say, but it certainly wasn't the last.

For generations, several mothers of kings were as ambitious and ruthless as Lady Macbeth, wielding enormous influence in affairs of state. In the fourteenth and fifteenth centuries, women were allowed to own property, a rarity anywhere in the world at the time.[18] Still later, women voiced criticism of the social order in long narrative poems that were taken very seriously. Some observers even suggest that the relative success that Catholic missionaries had in Vietnam is partly due to the way that version of Christianity values the Virgin Mary: the idea of a very powerful woman associated with God was a great fit with venerable Vietnamese ideas.[19] The young bride dancing on the roof of that bunker at the top of Hải Vân Pass might be seen as just the latest in a string of assertive women.

Be that as it may, at the end of the first millennium the Vietnamese won their independence from the Chinese in a decisive battle that took place near what is today the port of Haiphong, where the Red River forms an estuary. At that point the ruling Chinese dynasty was falling apart, and Chinese leaders were busy with conflicts on the many borders of their empire. Rumours that the Viets planned to take advantage of that weakness to revolt again led the emperor to send a large fleet of ships south to the mouth of the Red River.

A crafty Viet general lured them up the river at high tide, and then attacked when the ships were well away from open water. Surprised, the Chinese commander ordered a retreat back down the river, but the tide had turned. The ships were all impaled on iron-tipped stakes exposed by receding water.

The story of the brilliant trap was told again and again by the Viets, but it seems that it was forgotten by the Chinese. Two hundred years later the forces of Kublai Khan, the Mongol emperor who then ruled China, were caught by the same ruse at the same place. Seven centuries afterwards, American soldiers routinely faced a similar kind of trap: witness those short films showing grunts slogging away through rice paddies, and coming on sharpened bamboo spikes hidden in the water. As a song from the period put it: "I've got the Viet-Cong blues / Got bamboo spikes in my tennis shoes."[20]

Once free from direct rule from China, in the eleventh century the Vietnamese consolidated their position to the north of the passes. That done, they turned south to claim land for their burgeoning population, only to find another empire south of the Hải Vân Pass. Called the Cam, or Cham, the founders of this civilization appear to have moved into the region by the second or third century CE, and they prospered there for nearly a thousand years. Marco Polo, among others, was impressed by them: "a very rich country of wide extent," he wrote of the Cham kingdom after his visit in about 1285.[21]

Visit the temple complex the Chams built at Mỹ Sơn near Da Nang, in central Vietnam, today and you'll see a dim echo of their accomplishment.[22] It is perhaps the oldest inhabited archeological site in Vietnam, but much of it was destroyed by American carpet bombing in 1969. Of all the destruction wreaked during the various wars fought in Vietnam, this is among the most gratuitous. Footage exists of B-52s laying down bombs—the ungainly behemoth of an airplane opens its bays and dozens of bombs fall almost gracefully to the forests below.[23] Then comes the explosions, the smoke, and—especially when the bombs were filled with napalm—fire.[24] What's left today at Mỹ Sơn is nonetheless impressive: more than seventy temples and tombs, some dating back to 380 CE. This site, as well as other smaller ones in central Vietnam, bear witness to the vigour of the Champa civilization, which appears to have been quite different from both the Viet and Chinese ones. Their language used an alphabet derived from Sanskrit, and at the height of their power they were Hindu, one of many belief systems that people on the peninsula have followed over time.

Religion, as we'll see several times in our consideration of unidentical twins, can be an enormous force, both for creating identity and for consolidating influence. On the Southeast Asian peninsula, like everywhere else in the world, the first religion was almost surely a faith in the special powers of animals, specific places, ancestors, and what might be called hometown heroes. But religions with a purely local base—for example the ones that said a particular spring was sacred, and that the mountain over yonder was home to a powerful god—began to be pushed aside in the fifth and sixth centuries BCE. That was when a number of religious movements and ethical systems with a much wider scope were born almost simultaneously in widely separated regions. The founders of four of these movements—Zoroaster, Confucius, Gautama Buddha, and Lao Tse—all preached within roughly the same two-hundred-year period, while at about the same time the great texts of Hinduism were being written down. Some religions—like Hinduism, the oldest among them—did not seek converts, but because their adherents travelled, traded, and colonized, they spread their faith to places like Champa. Others either sent out messengers—as did the Buddhists—or were imposed by imperial decree—as Confucianism was on the Viets by the Chinese. The latter faith played an important role in Vietnam's social, political, and educational life, but most scholars agree that it was more a belief system for the elite while ordinary folk were more attracted by indigenous religions or Buddhism. It should be noted that Confucianism's emphasis on filial piety and respect for authority in a centralized state made it a very useful tool for keeping order.[25]

Champa statue of a dancer from the ninth to eleventh century. The Champa empire blocked Vietnamese expansion south for several hundred years. *Source: Daderot, Creative Commons 1.0 Universal Public Domain.*

Two monotheistic religions came to Southeast Asia considerably later. Muslim traders brought Islam to the Champa territory during the same period that the faith spread to Malaysia and Indonesia: the surviving Chams in Vietnam and Cambodia today are mostly Muslim. Christianity, particularly Roman Catholicism, became important in the political as well as the spiritual life of the Vietnamese, as we shall see.

Let us return again, though, to the turn of the first millennium CE. The Cham city states were strong and rich just at the time that the Viets, having thrown off direct Chinese rule, were ready to affirm their existence and move south.[26] But the Chams fought back, and it wasn't until the eighteenth century that the Viets finally vanquished them. That set the stage for Vietnamese expansion farther south into the peninsula. Two poles of power subsequently developed in Vietnam, one based in the old Viet heartland in the North, and the other in the South, in the Mekong Delta, with Saigon as its major city.[27] At the same time, perhaps somewhat ironically, the stage was also being set for the Vietnamese, who in effect were successfully colonizing much of the peninsula, to be colonized by Western powers.

There had been some contact between the peoples of Southeast Asia and Europe for centuries but it had been mostly overland: while Marco Polo travelled by sea during his voyages, a good part of his journey was made on foot or by donkey or camel. But travelling by ship is frequently faster, and, besides, you can carry far more in a ship than you can with even a long string of pack animals. So the European desire for the riches of the East led to attempts, beginning in the fifteenth century, to find a sea route there.

Every North American school child learns about Christopher Columbus's westward voyage in 1492, when he inadvertently "discovered" the Western Hemisphere for the Spanish Crown as he sought a shortcut to Asia by sailing west. But at the same time the Portuguese were headed toward the riches of the Indies by sailing south and then eastward around Africa. Roman Catholic missionaries went with them: Saint Francis Xavier reached Japan in 1549, with the first Jesuits arriving in Vietnam sixty-six years later.[28] By 1651, contact was advanced enough that a Portuguese-Vietnamese-Latin dictionary was published in Rome. It was the work of French Jesuit Alexandre de Rhodes, and used the Roman alphabet with added diacritical marks. Now called Quoc-ngu (national language), the alphabet was first used in Christian communities; by the late nineteenth century, however, it was taken up by nationalist movements, since its thirty-seven letters made reading much easier than either the Chinese and Nom systems, which required the memorization of thousands of characters.[29]

As trade routes between Asia and Europe developed in the sixteenth and seventeenth centuries, Saigon, located near the tip of the Indochinese Peninsula and the mouth of the Mekong River, prospered. Ports along the east coast also became more active. But Hanoi and its accompanying port of Haiphong suffered because, tucked away as they were near the head of the Gulf of Tonkin, they were out of the range of shipping traffic, which increasingly headed for ports farther east on the Chinese coast and to Japan.

This was the time when one dynasty, the Ly, was supposed to be ruling the entire country, but it was really governed by two, quite different, families. In the north the Trai branch attempted to rule, but did so badly. In the south, the Nguyen branch gave lip service to the northern rulers while becoming more and more independent, until it was clear that in all but name Vietnam (then called Dai Vet) was two countries.

It was only after a rebellion by three brothers toward the end of the eighteenth century that the country was united for a while. Promising land reforms, equal rights, and liberty at a time when the winds of change were also blowing through Enlightenment Europe, the middle brother declared himself emperor after successfully putting down another attempt at invasion from China that had been instigated by his opponents in Hanoi. But he died after only a few years on the throne. More disorder followed until one of the few aristocrats left from the previous period successfully took the throne, bargaining for help from the French. In exchange for trade concessions and some promises of religious freedom, he got support from French artillery.

With that the die was cast. The repercussions of this diplomatic horse trade can still be felt today.

At the turn of the nineteenth century, about 10 percent of the population in the southern part of what is now Vietnam was Christian.[30] Many of them lived in the Mekong Delta: K.W. Taylor argues that this is because there was greater freedom in this culturally diverse region where Khmers, Chams, and Chinese traders mingled daily. (The effects of people of different kinds regularly rubbing shoulders is something we'll return to later in the chapters about Alberta and Saskatchewan and the United States and Canada.) Nevertheless, there was official condemnation of Roman Catholicism in some powerful quarters, and over a period of some fifty years, thousands of Catholic clergy and communicants were put to death. Many of their names are unknown, so in 1988 Pope John Paul II decided to canonize a representative group of 117 Martyrs, including priests, nuns, and Catholic laymen and -women.[31] The French used this religious oppression as a pretext to attack and to gain concessions: after their defeat at home at the hands of the Prussians in 1871, they were particularly interested in finding compensatory prestige and influence.

The French had also been looking for a place to use as a base for trade and to balance British influence in India and Hong Kong. They hoped to control the Red River, which they thought was an important route into the Chinese province of Yunnan, but they also had their eyes on the country's southern ports. Rather than deal with a central government, in the end they found it easier to make arrangements with separate parts of the peninsula. Thus, by the end of the nineteenth century France had signed several agreements that institutionalized a French Indochina consisting of five provinces: Tonkin (in the North, in essence the Red River Valley,) Annam (including the coastal land south of the Red River Delta, much of which had been Champa territory,) and Cochinchina (in the South curving around to embrace much of the Mekong Delta). (Laos and Cambodia were the other two territories.[32])

In the twentieth century, French influence worked its way into the fabric of Vietnamese society. Sons of the Viet elite—including a scion of the ruling Nguyen family—were educated by Catholic missionaries. Hanoi, the capital of French Indochina, was built "as stately as a precinct of Paris," in the words of the American writer Paul Theroux.[33] Saigon was also a bustling port, the hub of France's *mission civilisatrice*, that country's high-minded justification for trying to remake Indochina in its own image.

The French connection initially sheltered Indochina from some of the worst treatment by the Japanese early in World War II. After the German invasion of France in 1940, Indochina was ruled by the Vichy government in France. For a while the Japanese were more or less content to leave the pro-Axis French in control, but in March 1945 they took over, imprisoning all French administrators. When the end of the war came that summer, the Japanese in the southern province of Cochinchina surrendered to British forces—Gurkhas who had been fighting in neighbouring Burma—since there were no French military available to accept the Japanese recognition of defeat.[34]

In the North, the Allies' plan had been for the Japanese to surrender to the Chinese forces of Chiang Chieh-shih, but before that could happen the nascent Vietnamese nationalist movement under Ho Chi Minh declared the country's independence from the French. His proclamation contained words that might have given hope to those who wanted to see an independent democratic state on the Southeast Asian peninsula: "All men are created equal. The Creator has given us certain inalienable rights: the right to Life, the right to be Free, and the right to achieve Happiness....These immortal words are taken from the Declaration of Independence of the United States of America in 1776."[35]

High-minded phrases, which were not honoured. When the dust settled, Ho's Viet Minh forces took over the territory north of the 16th parallel—the country that had been the ancestral home of the Vietnamese. In the South, rival groups vied for control until a French expeditionary force arrived, ready to reimpose French colonial order. By the end of 1946, the French had announced they were going to take control back in the North too.

The result was what the Vietnamese call the First Vietnamese War, bringing with it the struggles of the Street Without Joy. It was a classic guerilla conflict, with the Viet Minh able to melt into the countryside when challenged by the initially better-armed colonial forces. The war dragged on until 1954. Then the French, who were facing challenges to their rule elsewhere in their colonies—as we shall see when we look at Algeria and Tunisia—withdrew, following a crushing defeat when their fortress at Dien Biên Phu was besieged by Ho Chi Minh's forces.

The peace accord called for large movements of the population from one part of the peninsula to the other. At least 800,000 people, mostly Catholics and members of ethnic groups favourable to the French, moved south, while 80,000 others sympathetic to Ho Chi Minh's cause moved north. This population shift reinforced the differences between the North and the South. So did the outside forces that came in aid of the two sides, China in the North and the United States in the South.

— — —

Which takes us to the summer when my husband and I came to Canada and the United States was embroiled in a war that seemed to have no end. We left the San Francisco Bay Area about noon on Wednesday, August 28, 1968, our vw Beetle loaded with most of our worldly possessions. By Friday night we'd made it to North Platte, Nebraska; the plan was to stop on Saturday somewhere around Chicago, more than seven hundred miles to the east. But we hadn't figured on the Democratic National Convention, which wound up that day. Norman Mailer called it the "Siege of Chicago" in his account of the presidential nominating conventions that year, and hordes of people had flocked to the city to participate and protest. Getting a place to stay in the Windy City was going to be impossible, we figured, so we went off the highway and into the suburbs. But in that time long before the Internet—young folks often didn't even telephone to book a room because long distance calls cost so much—we had to rely on spotting "vacancy" signs at motels.

No luck. The combination of the convention and Labour Day weekend meant that even in Joliet, forty miles outside of Chicago, we couldn't find a motel, however sleazy. There was nothing to do but push on that night.

As a result we arrived in Detroit early Sunday morning after a couple of hours of sleep at a roadside rest stop, and more than a thousand miles from where we had started twenty-four hours before. With our hearts in our hands, we crossed the border into Ontario at Windsor, not sure what to expect because a postal strike meant our documents weren't complete. There also was the question of my husband's draft status. At the time he had a student deferment, and even though the Canadian government had said that draft status was not to be considered in evaluating immigrants, we were worried. But after a short interview, a quick calculation of our place on a "new immigrant" scale, and presentation of a letter from my husband's future employer, the immigration official shook hands with us. "Welcome to Canada," he said.

Okay, we told each other, we've arrived. We relaxed, then pushed on, arriving in Montreal on a hot, sultry Labour Day. Reflection on why this country was different from the other could wait. So could wondering why there were two Vietnams.

— — —

Two months after we arrived in Montreal, Republican Richard Nixon handily beat Democrat Hubert Humphrey in the American presidential race. The anti-war movement was not pleased, but in retrospect it seems that protests did have an impact. Nixon, who clearly wanted to open up relations with China, and who would visit that country in 1972, began turning the fighting over to the South Vietnamese and removing American troops from Vietnam. Peace talks had already begun in Paris and would continue for four more years. The war went on, however, in part because the Americans insisted on having prisoners of war released before stepping out of the picture completely. It wasn't until January 1973 that a peace agreement was signed: the United States was to reduce military aid to the South drastically, and eventually Vietnam was to be reunited. But in the meantime, no longer supported by American equipment, the South's army faltered, faced with North Vietnamese who were now backed by the Soviet Union. Saigon fell on April 30, 1975, in part because the Southeast Asian war, so hugely unpopular in the United States, had become merely an eddy in the global currents of power flowing between China, the Soviet Union, and the United States. There's a huge irony there, of course. The Domino Theory had it wrong: in the end the Chinese had no desire to have a unified, well-armed Vietnam on its southern border after all those centuries when the Viets had frequently been an irritant and sometimes a real threat.[36]

Although the country was united a year after Saigon fell, the new Vietnam was not at peace.[37] Struggles with Laos and Cambodia continued for several years. China also briefly invaded in 1979, and border conflicts persist to this day. In the transition, hundreds of thousands of people were displaced, going north or south or fleeing the country, leading to a diaspora of legendary size with effects on other countries that we'll consider later.

Hanoi is now the capital of the much more prosperous Vietnam, but regional differences remain. Some are as simple as what a family puts out as offerings on its home altar during the Têt Lunar New Year celebration. In the South five fruits are traditionally used: custard-apple or soursop (mãng cầu), coconut (dừa), papaya (đu đủ), and mango (xoài), since they sound like cầu vừa đủ xài, which means "we pray for enough money to spend" in the southern dialect. In the North an upturned hand of green bananas is often the centre of the arrangement. The bananas may hold a round yellow pomelo with mandarin oranges, persimmons, and kumquats clustered around: the emphasis is on abundance and colour, with no linguistic games involved.[38]

Other, more profound differences continue, arising directly from the country's deep past. As K.W. Walker noted as recently as 2010, "Northerners are more disciplined to accept and to exercise government authority, they are proud of inhabiting what they view as the center of Vietnamese culture, they tend to be cautious about contact with the overseas world, and they are inclined to view what is happening in China as a model. Southerners are more individualistic, egalitarian, entrepreneurial, interested in wealth more than authority...open to the outside world, and wary of how things are done in China."[39]

Yet the exodus of more than a million Vietnamese in the 1970s and '80s might have had the effect of dampening regional differences. Many of the Việt Kiều (overseas Vietnamese) have prospered elsewhere, sending back $13.781 billion in 2017, a not inconsiderable contribution to a growing economy.[40] While Vietnam has not been able to use their talents at home, the absence of these people's voices may have had a profound effect on the country. Many of those who left were allied with or fought for the South Vietnamese cause, and fled after Saigon fell and the North Vietnamese forces triumphed. Those who remained are far more likely to be accepting of Vietnam's modernized Communism, and more comfortable with a united Vietnam. As we shall see as we go along, this self-selection has an enormous effect on why places are alike or different, and why borders between neighbouring peoples so often persist.

ALGERIA
AND TUNISIA

Let us imagine a bright blue day on the Mediterranean a long time ago, at the beginning of another diaspora. The ship, propelled by one sail and tens of oarsmen, has crossed from one of the islands in the middle of the sea, slipping out of port at night on the evening winds. The destination had been just over the horizon at that point, but now, in the morning sun, the hills protecting a harbour have become visible. Seabirds wheel in the wind, seeking fish in the bountiful waters. The land itself appears dark green, golden brown, and dusky purple. The voyagers are both anxious and exultant.

They are led by a woman called Elissa in some accounts, Dido in others. She was the daughter of the king of Tyre, who had willed her half of his kingdom on the eastern edge of the Mediterranean.[1] The other half was to go to her brother, who was enraged by the idea and plotted to take the crown for himself alone. He assassinated Elissa's husband, and would have done the same to her if she hadn't escaped. After a stay on the island of Cyprus, she set out for the African coast to establish a new city and base of power. With her on this morning are some loyal retainers and eighty young women destined to become sacred prostitutes, but who in the new land would become the mothers of Carthage, one of the greatest and longest-lasting city states in history.

The year is somewhere around 830 BCE, and Elissa and her companions are not really going into the unknown. Phoenicians traders—and Tyre was a Phoenician city, although at the time there was no such thing as a Phoenician empire—had already established a trading post at the mouth of a river now known as the Medjerda, on a promontory that protects a relatively safe harbour in Tunisia. All trade between the western and the eastern Mediterranean has to pass offshore: the strait between the North African coast and the island of Sicily is about two hundred kilometres wide at this point. It would be a great place to build a new city (and Carthage, or *Qart-Hadasht*, means just that in the Phoenicians' language, Punic). But to do so Elissa must get permission from the local population.

They are not fools, you can be sure, although we know little about them except that they are reluctant to grant permission. Once landed on the beach, Elissa must argue with them. In the end she persuades them to give her just enough land that the hide of a cow can cover. So a hide is produced while, one imagines, the Phoenicians' ship waits at anchor just offshore and the locals stand around thinking they've got the better of the deal. But Elissa proceeds to cut the hide into a lattice work so fine that when it is extended it covers what amounts to a square mile.

Like indigenous populations again and again, the locals have been had.[2] The sun beats down, the seabirds cry, and history takes a lurch in another direction.

— — —

Flash-forward nearly three thousand years to another city on a sheltered harbour where the sun shines down and seabirds cry. The teaching assistant in the introductory political science course I took in my second year at the University of California at Berkeley was blond, short, a little pudgy, and always wore a coat and tie. "Square," we said as soon as we saw him. Far too clean-cut to be cool. He looked like he didn't have a clue.

But it was he who shook the foundation of my core beliefs. Like most young Americans of my generation, I was profoundly ignorant. For most of us, it seemed we had been born in the greatest nation the world had ever known at a time when everything could only become more prosperous, more humane, more...well, just better. In part we were right. The United States at that point was in an exceptional period when the rich were "few in number and, relative to the prosperous middle, not all that rich," as economist and Nobel laureate Paul Krugman puts it.[3] There was much that was wrong, which many of us didn't see at that point; we had no idea that civil rights struggles at home and the Vietnam War abroad would soon open up the fault lines of the society. That October morning when the TA astonished us by showing up unshaven and unkempt counts as the beginning of my coming of age.

He was brandishing a week-old copy of *Le Monde* that had just arrived by air in that pre-Internet era of slow news transmission. There'd been riots in Paris that hadn't been reported in North America, he said: Algerians were protesting the way that the French government was repressing the independence movement in France's North African colony. Police had bashed heads, broken legs and arms, and hundreds had been held without food or water for twenty-four hours. There'd been torture, deaths. And back in Algeria, he said, things were worse. His voice breaking, he told stories of men with electrodes attached to their genitals, and girls who had light bulbs forced up their vaginas. I remember being shocked: I don't think I had ever heard any man say "vagina," and the idea of violation by dangerous object was something I, as innocent sexually as I was politically, could not fathom.

We had to take his word for it—none of us could read the newspaper because none of us knew French—and the import of what he was telling us did not become clear until much later,[4] just as the number of deaths in those riots in Paris was not revealed for a couple of decades.[5] (Note that Anglophone Canadians of this generation would have been, even then, far more likely to know enough French to read a newspaper, even if it was hard to get one.) Today it is generally believed that somewhere between forty and two hundred young Algerian men were killed that October day and dumped in the Seine. France, which had experienced the sorrow and humiliation of the Street Without Joy in Vietnam, was fighting savagely in Algeria in hopes that it would not have to give up this rich North African colony.

Nevertheless the French were forced to do just that, and in March 1962 Algeria became independent. There followed fifteen years of civil war, which damaged the country even more.

The connection between these horrible events and the founding of Carthage is not readily apparent, but they are linked by the themes of colonialism and greed.

Phoenicians controlled the northern, Mediterranean coast of Africa for nearly a thousand years. Then they were displaced by Romans, Arabs, Vandals, and Turks, each in succession: the French were really Johnny-come-latelies.

By the time they were fighting to retain their colonies, North Africa had been divided into countries and territories that share much history but whose paths have diverged.

Algeria's story is not unique among the nations born during the tumultuous decade of the 1960s, which saw at least two dozen independent countries come into being in Africa, free from colonial bondage. Fifty years later, the optimism and promise has morphed into despair and corruption in many of them. Grand dreams of free, just, prosperous nations have been subverted.

Yet, I discovered when I began considering places that are alike yet different, just next door to Algeria is a country that is following a much different path in the twenty-first century. That country, Tunisia, was where the Arab Spring of 2011 began, and is today the only country where that popular uprising led to substantial improvements. A quartet of Tunisian civil-society leaders won the Nobel Peace Prize in 2015. There are problems—some of them severe—but Tunisians so far have succeeded in hanging on to most of the advances they won in that hopeful spring.

Why?

To find the answer, start by taking a look at the map of Africa—Google Maps does nicely because it shows patterns of vegetation revealed in satellite photos. The Sahara Desert covers about one-quarter of the African continent, from the Red Sea to the Atlantic Ocean. To its north, mountains and the coastal plain along the Mediterranean are watered by winter rains: a swath of green across Algeria and Tunisia shows the most productive land.

Anatomically modern human remains found in Morocco have recently been dated to around 300,000 years, making them the oldest evidence of folks like us found anywhere.[6] But there's a big gap in the fossil record lasting millennia.[7] Currently the record in North Africa picks up again 40,000 years ago or so, with strong indications that people were living in hunting-and-gathering societies.[8] By 10,000 years ago we can see a more complete picture in several sites dated by Carbon-14 and other processes. Such a long period of habitation suggests that there was plenty to eat in the region—among the finds are literally millions of shells from large land snails that have been burned and crushed, as well as the remains of butchered sheep and other good-sized animals.[9]

The people who lived in North Africa had a sophisticated tool kit of knives and other blades, and, it appears, a belief system that encouraged careful burying of the dead. Their culture and way of life were in many ways very complex. As early as twenty-five hundred years ago, one group, the Garamantes, discovered a way to tap underground water in what is now Libya, to the south and east of Tunisia.[10] They and their slaves bored tunnels deep into mountains that flanked the valley where they lived in order to reach the water. Ongoing archeological excavation, coupled with analysis of satellite photos, indicate that they were able to live comfortably for at least six hundred years before the water ran out.[11]

But about the time that Chinese civilization began to flower, the pastoral and nomadic people indigenous to the coastal plain were confronted by newcomers from the eastern end of the Mediterranean, from what is sometimes called the Levant. (That term comes from the French verb *se lever*, to rise, and is an allusion to where the sun rises—the east. It resonates wonderfully with an Arabic name for North Africa, Maghreb, which refers to the place where the sun sets—the west.)

The invaders, the Phoenicians, were seafaring traders looking for markets, and the first foreigners we have records of who wanted to control the southern coast of the Mediterranean. At first, most of the Phoenicians' settlements were small, places where goods could be exchanged with the locals, and raw materials like silver prepared for shipping to larger centres. But when Tyre, the Phoenicians' greatest city in the Levant, was besieged for years by

The Phoenicians founded the city state of Carthage around 800 BCE. It was razed by the Romans in 146 BCE. *Source: Patrick Giraud, Creative Commons Attribution-Share Alike 3.0 Unported licence.*

the armies of that ancient powerhouse Assyria, they set out to develop their overseas bases more extensively. This is where the myth of Elissa founding Carthage began.

Today the traveller from abroad is likely to arrive in Tunisia at the Tunis-Carthage International Airport, slightly to the north and west of the UNESCO World Heritage Site where the ruins of the great city are preserved.[12] They are eloquent testimony to the more than six hundred years that Carthage was first a military as well as a trading power—it was from here that Hannibal set out to conquer Rome with his men and elephants—and then was reduced to rubble by the Romans, who inevitably came into conflict with the Carthaginians as their territorial ambitions grew.[13] There were treaties and broken promises between the two powers over four centuries and, finally, three massive Roman mobilizations, the Punic Wars. "Carthage must be destroyed" became a watch word in Rome—one senator ended every speech he gave with that exhortation—until finally it was, after a siege that lasted three years and saw as many as 200,000 people killed.

It was a genocide, some modern historians say, an entire people killed for reasons that would today be called racist, but which were economic and cultural as well.[14] Sadly, we have few texts telling the Phoenicians' side of the story, since most of their writings, if they weren't destroyed by their enemies, have been lost because they wrote on papyrus and not the less perishable clay tablets used by many other early civilizations. What has survived is largely negative, like Homer's comment in *The Odyssey*: "Thither came Phoenicians, men famed for their ships; greedy knaves bringing countless trinkets."

There's a huge irony here, though. Homer's great epic was not written down until centuries after its composition because the Greeks, which at one point had an early script, lost it or cast it aside during years of drought, famine, and conflict around the turn of the first millennium BCE. *The Iliad* and *The Odyssey*, which are about wars and adventures that took place during that period, were handed down from storyteller to storyteller for a good three hundred years until *phoinikeia grammata*, or Phoenician letters, were introduced.[15]

Yes, Phoenician writing, to which the English text you are reading is intimately linked and which is perhaps the most enduring accomplishment of the Phoenicians. They had the great idea of assigning the twenty-two sounds used in their language to each of twenty-two symbols, adapted from Egyptian hieroglyphics, which, like the Chinese writing system, had been based on pictograms representing an idea. An example is the character that stood for "house" or "bet," which the Phoenicians associated with a "B" sound. With that process in mind, it was a relatively easy matter to make correspondences between symbols and sounds that reflected what the language sounded like and that didn't require memorizing thousands of hieroglyphics or characters. Furthermore, the script, other peoples discovered, could be adapted to any language, with a change here or there. Subsequently a clutch of alphabets were created, some of which, like those used to write Aramaic, Arabic, and Hebrew, are close to the earliest *phoinikeia grammata*. Linguists discern a straight line uniting Latin, Cyrillic, and Greek scripts, while modern Vietnamese script and Chinese *pinyin* writing demonstrate the marvelous flexibility of the idea of one letter = one sound. The curvy alphabets used in India to write Dravidian languages like Tamil and Malayalam, which we'll discuss in the next chapter, also appear to be derived either directly from Phoenician writing or, somewhat later, from Aramaic.[16]

The pattern of groups of dissatisfied people leaving Phoenician cities was common, it seems, and became policy in Carthage. Aristotle comments on it, saying that despite the oligarchical nature of government in Carthage, Carthaginians were "most successful in avoiding civil disturbance by sending out from time to time a certain number of the common people to their subject States and thereby enabling them to acquire riches. This is their means of healing the wounds of the polity and placing it on a permanent basis."[17] Many other powers have since followed suit, implicitly or explicitly, including Fidel Castro, who encouraged malcontents to leave Cuba in 1980.[18] It's a theme we'll return to again in our discussion of places that are alike yet not alike. It's also interesting that a woman gets such a prominent place in Carthage's founding myth: the place of women in a society can be an indicator of how two similar places go in different directions.

— — —

When the Phoenicians arrived on the coast they encountered a vigorous culture, and indeed it was still there when the Romans took the Phoenicians' place as rulers of the southern Mediterranean. In many respects the culture continues today among the Berbers.

For many North Americans of a certain age, say "Berber" and what springs to mind is the image of gallant, handsome horsemen charging out of the desert in Hollywood action films like *The Wind and the Lion*. But the various Berber clans were—and are—much more. Back when the Phoenicians arrived, many

Berbers lived south of the coastal plain as herders moving from place to place with the seasons in order to change pastures, the act of which in Greek was nomados (νομαδος). So they came to be called "nomads", and their kingdom, called Numidia, stretched over much of North Africa between the mountains and the Sahara. Speaking languages unknown to Greeks and Romans, they were also called "barbarians," the name that Greeks and Romans applied to any foreign people whose speech sounded, supposedly, like a jumble of sounds, or bar bar. By extension, the population of much of North Africa came to be called Berber. Today the name is a hot button with erstwhile Berbers, particularly in Algeria and neighbouring Morocco: they call themselves i-Mazighen, which means "free people" or "noblemen" in their language.

Later, during part of the Ottoman Empire (from about 1600 to 1850), the whole North African coastal plain was known to Europeans as the Barbary Coast. As long as they received a percentage of the haul, the Ottoman rulers turned a blind eye to what were called Barbary pirates, who attacked shipping from the many small ports. Victims considered the resulting damage so serious, however, that other countries went to "war" in attempts to stop it. Indeed, the first American military adventure outside North America took place during the Barbary Wars between 1801 and 1815. The clashes, which involved more than a dozen American ships, are immortalized today in the lyrics of the "Marines' Hymn," a ballad that is among the first evidence of American jingoism.[19]

But to return to the Roman conquest of Carthage: once the city was destroyed, the Romans lost little time in adapting its strategic position to their own ends. Under Augustus Caesar they rebuilt the city as the centre of the Roman Empire's African domains. The indigenous population was co-opted, but the Romans also brought in colonists from elsewhere in their empire.[20] The idea that the Romans could "own" the region conflicted with the ideas of the nomadic tribes about what territory they could move across and control, very much as Indigenous people who followed migrating bison did not acknowledge the boundary line between the United States and Canada in the mid-nineteenth century.

What is clear is that if the boundaries of Roman influence along the shores of the Mediterranean had been the blueprint for Algeria and Tunisia when they became independent, modern states in the twentieth century, the two countries' geographical size would now be more equal: modern Tunisia is about the same size as the part of modern Algeria that had been directly governed by the Romans. But Algeria today is physically much larger than its coastal population corridor. It includes much more Berber territory, lands where the nomadic culture was strong. That fact has immense implications for how the country developed, and why it is different from Tunisia.

— — —

The spring before my husband and I took our road trip from Berkeley to Montreal the award-winning film *The Battle of Algiers* was released in North America. Made by Italian director Gillo Pontecorvo, the film is a mostly sympathetic account of what happened when the Front de libération nationale (FLN) took the independence struggle into the city of Algiers in 1953.[21] It resonated in North America. As film critic Roger Ebert commented, "those not interested in Algeria may substitute another war."[22] Certainly that other war raging in Vietnam was on our minds when we saw it at a showing organized by the Film Society at McGill University in Montreal. I remember being mightily impressed, and thinking of what that TA in Berkeley had tried to tell us years before.[23]

Algerian women traditionally wore the *haïk*, with a triangular face scarf, the *aadjar*, but during the struggle for independence some put it aside, the better to fight.
Source: Mustapha Brahim Djelloul, Creative Commons Attribution-Share Alike 4.0 International licence.

One sequence, in particular, has stayed with me. In it, three women independence fighters, who until then had always worn the traditional Algerian white robe, the *haïk*, with a triangular face scarf, the *aadjar*, in public, prepare to dress like French women so they can circulate freely in the Europeanized neighbourhoods of Algiers. The camera shows them appearing to be reluctant to uncover their hair. But they do, styling it fashionably: one even gives hers a peroxide treatment to make it lighter. Then, dressed in knee-length skirts and blouses with short sleeves, they leave their quarter carrying the makings of bombs in their up-to-date handbags. (I had a white wicker purse like that carried by one of the girls.) French soldiers at checkpoints don't question them, and twice flirt a little. The women proceed to crowded places—an upscale bar frequented by a crowd of successful, middle-aged people; a cabaret where young people cha-cha (hopefully? ironically?) to "Hasta mañana"; the local Air France office. In each case they look around at the crowd a little hesitantly before leaving their purses and then slipping away: it's clear that their actions are taken in full recognition of the consequences, of the destruction they will cause when, moments later, their bombs go off.

Viewed fifty years later, at a time when news of terrorist attacks is splashed across the Internet nearly every week, my feelings are mixed. Suicide bombers, assault rifles, heavy vehicles crashing into crowds: nobody wants them. But the courage of these women and their devotion to their cause seem to me understandable. Sadly, though, their audacity and that of other women freedom fighters did not translate into an important place for women in Algerian society after independence. It took twenty years before an Algerian Family Code was passed, and, rather than validate the place of women, it appealed to Islamic fundamentalist forces in Algeria by adhering closely to conservative elements of sharia law.[24] Only in 2015 was violence within a conjugal relation prohibited.[25]

In contrast, Tunisia's Code of Personal Status was enacted five months after independence in 1956. It abolished polygamy and a man's ability to divorce his wife by simply repudiating her. In addition to giving women the right to ask for divorce, the law also decreed that both spouses must consent to marriage and be eighteen or older. Tunisian women got the right to vote in 1957, and to seek election to office in 1959. The Tunisian constitution also guarantees the principle of equality, which has led over the years to the entry of women into many non-traditional fields.[26]

It's tempting to say that better treatment for Tunisian women goes all the way back to the times when Carthage was founded, and its crafty queen Elissa.

But since then many waves have pounded against the Tunisian shore, thousands of ships have anchored in the port's waters. More recent events have had a stronger impact, although there is a connection between the people who tried to keep Elissa from setting up a colony in their midst and the fate of Algeria and Tunisia today.

Islam has something to do with this, but not as much as non-Muslims might think. Both countries are overwhelmingly Muslim today, but in some respects that's a recent development in the region's long history. Rome was Christian by the time its empire collapsed, and much of the North African population was at least nominally Christian by then as well. Saint Augustine, the theologian and philosopher whose thought influences Christians still, was born in Thasgaste, or what is now Souk Ahras in Algeria, and was bishop of Hippo Regius, the modern Annaba. In addition there were several, even older Jewish communities. We'll see in the next chapter that after the fall of the Second Temple in 70 CE, some Jewish refugees fled east, as far as the western coast of India. But others went west. Perhaps as many as thirty thousand were deported to Carthage, where they joined other Jews who had been established there for three hundred years: the Jewish historian Josephus says that kings of Egypt settled Jews at Cyrenica, in Tunis, as far back as 312 BCE.[27]

Just how deep these religious allegiances were is unclear, but what is certain is that in the seventh century CE the region quite suddenly was swept by a philosophy coming out of the East, carried by martial forces inspired by the Prophet Mohammed. Islam changed everything "in the twinkling of an eye," as one historian puts it.[28] Within thirty years Muslims from the Arabian Peninsula controlled the world from the eastern shores of the Mediterranean, south of the Byzantine Empire, and as far east as Afghanistan and the Punjab in India. Then they turned west. The conquest of Egypt took just two years, from 639 to 641, while gaining control of North Africa took only a little while longer.

A measure of the success of an invading force is how much it transforms the ordinary life of the citizens they come to rule, particularly in what they believe and the language they speak. By that criteria, the Arabic Muslim invasions were among the most successful in history, certainly more than the Chinese ones in Vietnam, to say nothing of the French colonization of both Southeast Asia and North Africa. However, elements of the Berber populations of the North African hinterland, while they embraced Islam, did not adopt all of the new religion's teachings, including some strictures of the Quran that govern the most intimate parts of life, the family.[29]

The Prophet was revolutionary in his time because his teachings gave women the right to inherit, something that few cultures at the time accorded them. Today, traditional Muslim practice in this area may seem insufficient, cruelly unfair even—for example, daughters have a right to inherit only half of what their brothers can.[30] The idea of giving women full inheritance rights has elicited storms of protest even in Tunisia today.[31] But at the time Islam arrived, it represented an improvement since women in most clans in the Maghreb could inherit nothing. This "customary law" was in force for centuries.[32] Feminist scholar Mounira Charrad says that as late as the turn of the twentieth century, the French tried to subordinate it to French civil law but couldn't in what is now Algeria because customary law was too deeply embedded in local society. In Tunisia, though, the relatively more progressive Islamic tenets were honoured early on. Furthermore, by the nineteenth century, forces who wanted to push Tunisia farther toward modernity were able to liberalize the legal framework even more. In large part this difference between the two countries was due to two things: the conservative culture of the immense Algerian hinterland, and the sort of relation France had with its two colonies.

As noted in the previous chapter, the French officially saw their colonial activity as part of a *mission civilisatrice*, but their motivations were also commercial and inspired by a desire for power.

In North Africa, French involvement began in the 1830s with a trivial incident involving a French diplomat flicking an official of the then reigning Ottoman Empire with a fly whisk. A terrible insult! Looking for an excuse to act anyway, the French invaded.[33] After the dust settled, Algeria had been incorporated into the French nation, becoming a *département* like those found in continental France.

As it happened, French vineyards were suffering from an attack of a plant disease as deadly to them as potato blight had been in Ireland. The stage was set for making the hillsides of Algeria into the greatest wine-producing region in the world. Thousands of French vintners and wine workers crossed the Mediterranean ready to begin anew. By the turn of the twentieth century, a hundred thousand French had immigrated and by the middle of it, Algeria was divided into a ruling minority made up of Europeans and Europeanized locals governing a much larger, mostly Berber, mostly Muslim, population. (To be given full French citizenship rights, Algerian Muslims had to renounce their religion, and some did.)

— — —

I'm tempted to stop here to have a glass of wine while we consider what that means. Before we came to Montreal, my husband and I had never drunk anything but California wine, which at that time was just beginning to become more than rotgut. One of the pleasures of our new city, we quickly discovered, was having access to wines from many countries. Among them were sturdy, inexpensive wines from Algeria that we drank regularly. At the time—only a few years after Algerian independence—the strangeness of having so much wine from a Muslim country did not occur to me.

But the millions of barrels produced in Algeria were a measure of the changes that the French attempted to bring to North Africa, where the dominant religion forbade consuming alcohol. Not that the climate was all that hospitable to winemaking: it actually was too hot for yeast to ferment grape juice in the traditional way so a process involving refrigeration had to be developed. Yet within a few decades Algeria was exporting more wine than any other country.[34] That would change after independence: the winemaking industry was first nationalized and then dismantled as the country became more aggressively Muslim. Suffice to say that only one Algerian wine was available in Montreal when I was working on this chapter.

There were four Tunisian wines in Montreal wine stores, however, reflecting both the more secular nature of the country today, and its concerted attempt to expand its exports in the aftermath of the Arab Spring. That might seem surprising since Tunisia was never as close to France as Algeria, and so, one might think, less influenced by "progressive" European thought, and therefore less likely to develop a wine industry. But for several decades in the middle of the nineteenth century, something like an enduring Tunisian national identity had been established under Ottoman rule with the local *bey* attempting to modernize the country. He built railroads and created an educated elite, which understood both the ways of an Islamic state and modern European developments.[35] The effort was effective, but also expensive, and by 1869, Tunisia was forced to declare bankruptcy. The same sort of humbling had badly affected Haiti, as we'll see later, but here the result was political instability that the French were ready to turn to their advantage, invading with a 36,000-man force in response to a border skirmish with Algeria. While Tunisia became a French protectorate in 1881, the groundwork had already been laid for a quite different state than the one that would develop in Algeria.

When the winds of independence began to sweep through European colonies after World War II, the French made a calculation about their North African possessions: rather than fight on several fronts, they decided to concentrate on retaining control of the "territory that mattered most to it, Algeria."[36] Ten percent of Algeria's population, or nearly one million people, were European *colons* at the time, and were intent, political scientist Michael J. Willis says, to maintain *l'Algérie française* at all costs. When the war for independence ended, they felt threatened and left the newly independent Algeria en masse, 700,000 within the first year.[37] Today less than 1 percent of the population is classified as "European."[38]

In pre-independence Tunisia, Europeans—both French and a number of Italian nationals—made up a much smaller proportion of the population, and the visceral attachment to Europe was not nearly as strong on their part or on the part of the territory's French administrators. So in 1956, after seventy-five years of French rule, Tunisia rather peacefully gained its independence. Habib Bourguiba, who had studied law in France, took the reins of power, and enforced his vision of a modern Tunisia. Education was essential, he believed: universal primary schooling was achieved by the end of the 1960s, while the establishment of Tunis University was an early accomplishment. But the cornerstone of his policy was the Code of Personal Status, because Bourguiba believed that women's status and position were "demonstrative of the development and modernity of a society."[39]

In contrast, women in Algeria, who had played such a large role in the struggle for independence—as witnessed in *The Battle of Algiers*—were relegated to the sidelines. By the 1970s, while the economy became more dependent on oil production, a concerted effort was made to Arabicize education and to increase religious instruction, which further marginalized women. By the time dissatisfaction with economic and social conditions had mounted, producing riots in 1988, Islamist parties had become strong political players. Three years later, when it appeared that a coalition of Muslim groups, the Islamic Salvation Front, would win legislative elections, authorities cancelled the elections. There followed a civil war that lasted for more than ten years, during which as many as a hundred thousand people were killed.

As that war wound down, Abdelaziz Bouteflika was elected president in 1999 on a promise of good and stable government. But by 2014 good government had gone the way of the dodo, and despite suffering from the effects of a debilitating stroke Bouteflika was elected to the presidency a fourth time.[40] "Vote against change" was his slogan, coupled with a call to remember what had been wrought by the country's aging leaders when they were young: the end of 132 years of French rule. Despite his continuing illness, Bouteflika prepared to run for another five-year term in 2019, but changed course six weeks before scheduled elections when faced with massive, nationwide protests. He resigned, announcing he would not run again. New elections were scheduled for July 4, but peaceful protests continued.[41]

Tunisia escaped a civil war between Islamist and more modernizing forces, but unfortunately President Bourguiba also became an autocrat with the passing years, and did not step down gracefully. In 1987, when he was eighty-eight, doctors declared him unfit to rule, and Prime Minister Zine El Abidine Ben Ali took over. Ben Ali was re-elected several times with large majorities, but he, too, outstayed his welcome. By the first decade of the twenty-first century, he, his family, and his regime were generally perceived as corrupt, and not at all concerned about deteriorating economic conditions.

— — —

Compare and contrast: it took a week for that copy of *Le Monde*, with its news about Algerian protests in Paris, to get to Berkeley in 1961, but only half a day in December 2010 for videos of the beginning of the Arab Spring to make it to YouTube, and thence to the world.[42] In them, Mohamed Bouazizi, a young peddler of fruits and vegetables in the small town of Sidi Bouzad in the Tunisian interior, set himself afire. The aim of his desperate act was to draw attention to corruption and the country's astronomical unemployment rate, particularly among the educated young. Bouazizi had hoped for a university education himself, but he'd dropped out to help pay for the studies of his brothers and sisters, one of whom, Leila, was a third-year university student.[43] Police had confiscated the scales he used in his little business because he refused to pay a bribe. A policewoman had reportedly slapped him and insulted his dead father. Videos shot by his cousin and friends moments after he immolated himself show his pushcart surrounded by debris and first responders manhandling him to an emergency vehicle.[44]

By the time Bouazizi died of his burns two and a half weeks later, the conflagration he lit had spread across the Arab world. By then the Tunisian strongman Ben Ali had fled the country, and protesters in Egypt were demanding the ouster of the Egyptian dictator Hosni Mubarak. Shortly thereafter, the winds of change roared into Yemen, Libya, Bahrain, and Syria. The results of this Arab Spring, which began in peaceful protest, would be decidedly mixed.

Eight years later many observers considered the situation actually worse than it had been in Libya. Civil war in Yemen has seen the slaughter of some ten thousand civilians and the displacement of tens of thousands more, and in late 2018 the UN warned that the country faced a famine that could be the worst the world has known in a hundred years.[45] In Syria, where the dictator Bashir al-Asaad has tried to cling to power, the destruction has been almost beyond belief.

Tunisia and Algeria are exceptions to this, for very different reasons. In the latter it's because the Arab Spring never happened. Writing in the *New York Times*, the Algerian writer Kamel Daoud underlined the difference between

Demonstration in Tunisia during the Arab Spring of 2011. When Tunisian fruit peddler Mohamed Bouazizi set himself on fire in protest of government corruption, he ignited a blaze of protest that swept through the Arab world. *Source: M. Rais, public domain (original title of photo: Caravane de la libération 4).*

the two countries.[46] In Algeria, Daoud claims, Mohamed Bouazizi would have been bought off, because what protests took place there that spring did not call for democracy, but housing and roads, water and electricity. Daoud tells how, shortly after the Tunisian's death, another man set himself on fire in a town west of Algiers. "Reporters flocked to him, thinking they had found a revolutionary. 'I am no Bouazizi,' said the Algerian, from the hospital bed in which he would not die. 'I just want decent housing.'" In fact in 2011 some 130 young Algerians set themselves on fire in desperate individual protests against unemployment—which was then running at about 40 percent among the young—inadequate housing, food shortages, and police corruption. The result was not a wave of popular protest, like that which followed the Bouazizi suicide, because, Daoud asserts, Algerian President Bouteflika responded with just enough money for housing and promises of certain human rights reforms to stifle discontent. Besides, the Algerian Civil War still casts a long shadow in that country, making would-be protesters extremely wary of too much agitation.

Indeed, the long spasm of violence in Algeria in the 1990s can be seen to foreshadow what has happened in most of the countries convulsed by the Arab Spring: a period of euphoria following citizen protests, and then bloody conflict between forces that are more or less Islamist and those who believe that authoritarian government is the only answer to post-revolt chaos. That the latter stand to profit in many cases from control of a country's resources is an important factor in the equation, and may also influence the attitude of the rest of the world. Perhaps the big question is: Must an "Islamic Winter" inevitably follow the Arab Spring?[47]

Political demographer Richard Cincotta suggests that the answer to this question lies in the median age of the population. According to his research, there is a strong correlation everywhere around the world between a median age of at least twenty-six, civil peace, and movement toward liberal democracy. Cincotta forecast in 2008—well before the Arab Spring—that because Tunisia had reached that demographic milestone, it had very good chances of undergoing peaceful transition to liberal democracy. Political violence is more likely when populations are young, he argues, but as population structures mature, "citizens and the commercial and military elites are likely to question the need for an authoritarian government and reject the costs—in terms of civil liberties, political rights and corruption—that they bear."[48] By 2010 Tunisia's median age was 29.2 years, a population profile brought about by better health care and lower infant mortality coupled with lower birth rates.[49]

But correlation is not causality, and other factors were also at play. Before the Arab Spring a relatively strong civil society had grown up in Tunisia in the absence of the worst ravages of colonial or civil war. When a power vacuum developed in 2012, there were institutions whose leaders could claim some legitimacy and who were able to band together to draw a road map to reform and representative government. Set up in the summer of 2013, a coalition formed of the Tunisian General Labour Union (UGTT), the Tunisian Union of Industry, Trade and Handicrafts, the Tunisian Human Rights League, and the Tunisian Order of Lawyers was largely successful in bringing about consensus on how to proceed, work for which they won the Nobel Peace Prize in 2015, as mentioned earlier. Their example should give hope and guidance to other countries feeling their way toward peace and a democratic society, the Nobel committee said in their citation.[50] But there has been trouble in Tunisia since then, particularly over a lack of economic progress.[51] Most notably, in the winter of 2019 the UGTT held a nationwide strike over the government's refusal to raise salaries for civil servants, a measure that is part of a package of austerity measures demanded by the International Monetary Fund in return for financial backing.[52] The role of foreign financiers, both institutional and governmental, has been extremely important in determining the lot of poor countries, as we'll see when we look at other unidentical twins. But some still have hope...

— — · —

When I wrote this, the sky here was the intense blue of summer, and although we are far from the sea, the Saint Lawrence River was as blue as the waters off Tunisia have been since long before Elissa's ship made its way to the North African coast. Spring—Arab and otherwise—had passed, it was officially early summer, a lovely evening after a day of brilliant sunshine. What's more, it was an evening for celebration—Quebec's *fête nationale* as well as the end of that year's Ramadan, the Muslim holy month of fasting. Quebec liquor stores were well stocked with the four Tunisian wines and the one Algerian wine available here. Certainly some Tunisian-Canadians were going to include a little wine in their celebratory meals after the sun set and again during Eid al-Ftar, the feast days that followed. Some Algerian-Canadians would probably do the same.

So it was in a celebratory spirit that we sat down to do a little tasting of wines from both Algeria and Tunisia. We took notes: this one had not much "nose," that one's colour was a little thin, the third one had an earthy taste rather than the fruity notes promised on the label. The Algerian wine, we noted, had a sharp alcoholic taste: strange to think that a country where Islam is such a strong force would produce a wine like that.

But we came up with a winner: from Tunisia, a mixture of merlot and cabernet sauvignon grapes with a name that evoked the Phoenicians. It was from the hills above the River Medjerda, into whose mouth Elissa's ship had sailed. A link with the deep past, and a nod to the differences between two countries that are alike, but not really.

KERALA AND
TAMIL NADU

Blue sky, blue sea: you'll find both in abundance in India's two south-ernmost states, Kerala on the west coast and Tamil Nadu on the east. One is washed by the waters of the Arabian Sea, the other by those of the Bay of Bengal. Both lie only ten degrees or so north of the equator and the sun glitters off the water most days, while skies are blue for much of the year. They have much more in common, though. Frequently part of the same polity in the past, the languages their populations speak are close cousins, with roots going back deep into history and which boast an ancient literary heritage on a par with that of Greece and China. Today they are part of an Indian success story, yet they differ in ways that say worlds about what makes places similar and different, and about why and how places change.

About twelve years ago I visited Kerala when doing research for my book *Green City: People, Nature and Urban Places.*[1] My idea was to find an Indian city where the complex relation between people and the environment had taken some interesting—and perhaps instructive—twists and turns. A writer friend suggested Kochi, formerly known as Cochin, as a city where things were happening in a unique ecological system. The surrounding network of lakes, inlets, and estuaries is so beautiful and productive that it had been a stopping place for traders and adventurers for centuries, she said, and even today it was wonderful to visit. *God's own country* was the state's motto, she added with only a bit of irony.

Take a look at the map of Kerala and you might think you were looking at one of northern and central California: lowlands and valleys lying to the west of high mountains running north to south with a major port located on the estuary of several major rivers. Instead of the dry hills and valleys of the Golden State, however, Kerala is green with palm trees, rice paddies, and tree-shaded lagoons before rising to lush tea plantations in the foothills. Near the middle of the state the River Periyar, like the San Joaquin and Sacramento Rivers in California, fills a long, narrow bay before emptying into the sea.

Tamil Nadu lies on the other, eastern side of the mountains known as the Western Ghats: for the most part the border between the states runs in a sinuous north–south line along the height of the land. Some roads crossing the mountains give as many chills and thrills as Vietnam's famous Hải Vân Pass, although the views aren't quite as spectacular.[2] But once down, the landscape is drier than in Kerala, since the mountains act as a barrier to the rain-filled weather systems blowing in from the west. Tamil Nadu ordinarily receives most of its rainfall during the Northeast Monsoon, in October and November, and little during the Southwest Monsoon beginning in June, while Kerala receives rain during both periods.[3] The eastern state, however, profits from the abundant rainfall in Kerala, since several rivers cross Tamil Nadu after rising in the Western Ghats. The mountain range joins a lower, but still significant, one—the Eastern Ghats—to form the Nilgiri Hills in the south of Tamil Nadu, geographic features which have had a big impact on South India's history.

— — —

One of the first things that struck me about Kerala, and South India in general, was—and my notes repeat this several times—how handsome the people are. Almost everyone I met appeared slim, well-proportioned, most with shining white teeth, and all with the darkest skin I'd ever seen outside of Africa. The last was in sharp contrast to most of the South Asians I'd met in North America, whose paleness was partly due to living in a northern climate where the sun is so much weaker for much of the year: not much chance of getting a tan except in summer. But much was related to the fact most had their roots in North India, where the genetic mix is somewhat different from that of the South when it comes to skin colour. (Note, though, that recent studies show that all Indians share a similar genetic heritage.[4])

That difference—I hesitate to call it a "colour line," given the political freight that phrase carries, particularly in the United States—is the visible manifestation of population movements going back thousands of years. There are very good reasons why it's advantageous to have dark skin when you live in places nearer the equator, and lighter skin where days grow short and covering up with clothing is a necessity. In the former, dark skin filters out cancer-causing rays to some extent, while in the latter, light skin makes it easier for the body to manufacture Vitamin D, which is essential for proper bone development. Look on a map at where dark-skinned persons come from, and you'll see that in general they or their ancestors have their origins within fifteen degrees of the equator.

Despite this advantage when living in low latitudes, lighter skin is prized by many in India. Why is hard to say. Some would argue that the preference is linked to the fact that a series of lighter-skinned conquerors have invaded India from the north, and so light skin has long been associated with power. Indeed, one way of looking at the history of South Asia is to see wave after wave of conquerors coming from the north and trying to subjugate those who lived farther south.

The pattern of languages spoken in India corroborates this hypothesis. Northerners speak languages linked closely to Sanskrit and which are cousins of languages spoken in broad swaths of Iran, Afghanistan, and on into Europe. Southerners, however, speak languages that descend from Dravidian, which appears to once have been the language spoken much more widely in the Indian subcontinent but which was pushed aside by the invading northerners. Two other linguistic pockets of Dravidian languages remain on the edges of India, in what is now northwestern Pakistan and to the east, in the lower Ganges valley, but South India is the Dravidian homeland. As Irish travel writer Dervla Murphy noted in the 1970s, "there is considerably more resemblance between Hindi (originally envisioned as India's sole official language) and Irish than between Hindi and Tamil," the language of Tamil Nadu.[5] This linguistic difference and its corollary, the kinship between European languages and those spoken in North India, were first noted by European scholars in the early nineteenth century. Subsequently, it was used as fuel to bolster racist, pro-Aryan ideas denigrating the darker-skinned population of southern India.[6]

But Dravidian culture is ancient and rich. The Tamilakam (meaning "homeland of the Tamils") societies of the South were ones of great accomplishment. Their temples and monuments impress visitors today, and they left other cultural riches too. Among these is the body of Sangam poetry dating from a period of six or seven centuries before and after the beginning of the Common Era, when forms of Tamil were spoken both in Kerala and Tamil Nadu. Today, Malayalam is Kerala's language, and the difference between it and modern Tamil is large enough that someone who was raised there will have trouble understanding a native of Tamil Nadu, and vice versa, when each speaks his or her mother tongue. Yet until the twelfth century the two languages used the same script, developed from an earlier alphabet designed for writing on banana leaves: it is full of circles and swirls, which could be made without tearing the leaves.[7]

More than twenty-three hundred poems survive from the grand body of Sangam poetry. They were written by more than four hundred poets, men and women from many walks of life. This literature contains work of two sorts, one centred on the author's personal experience. The other is more concerned with heroism, valour, philanthropy, and ethics. Lost for more than a millennium, the poems were rediscovered in the nineteenth century, when they provided an impetus for the re-valourization of the culture rooted in the South. Certainly, their appeal transcends regional boundaries. The cycle *What the hero said to his charioteer* might be the song of a warrior from many societies, but, tellingly, it was written by an ancestor of the people who speak a Dravidian language today because their forefathers were able to repel invading enemies. An example:

Ride on charioteer! Ride your tall
chariot through the forest,
cutting into the red earth,
through the wide path
decorated with rain water,
passing dark blue *karuvilai*
flowers that look like sapphire gems,
bright *kānthal* flowers on bushes,
and bright golden *kondrai* flower
clusters that hang on every tree branch
like gold coins, spreading their fragrance.

Let us go to our fine house and listen
to the sweet words of my wife,
who loves being hospitable to guests,
the wide-shouldered young lady with
lightning-bright jewels, as she goes to
where our son with small,
unsteady steps and flower-like eyes
sleeps, to greet him as he wakes up.[8]

The names of several warrior kings are mentioned in the poems. Most of them are rulers of the Chera Empire (also known as the Keralputras). Centred in the west of what is now Tamil Nadu, its capital was on the extensive plain through which flows one of the tributaries of the Kaveri, a river that rises in the Western Ghats. At the beginning of the Common Era, the Cheras expanded toward the west coast through passes in the mountains and down the rivers draining toward the Arabian Sea. (The name Chera appears linked to a Tamil term meaning "declivity in the mountain," and is the origin of the name Kerala.)[9] The pepper growing in the green country west of the Ghats brought the Cheras great riches: the Greek geographer Strabo (64 BCE–24 CE) wrote that it was to Chera ports that the yearly expedition of as many as 120 ships from the Roman Empire sailed to bring back the spice.[10] But more about that later.

The theme of conquering and colonizing empires is one we've already encountered in this discussion of neighbouring states that are alike but not alike. It's also important to note that the first successful ones burst upon the world within two hundred years of each other, a very short time in the history of humanity. To be sure, groups have always quarrelled with their neighbours, but a concerted effort to subdue great regions was something new when Darius the Mede (550–486 BCE) brought the world from the mountains of Persia to the Mediterranean under his control. Less than a hundred years later Alexander the Great (430–354 BCE) swept his armies the other way, while in China regional rulers struggled to subdue each other. Then, a few decades before the Qin dynasty unified China between 221 and 206 BCE, the first great Indian empire, the Maurya, was established in the north of the subcontinent.

Chandragupta Maurya, the first Maurya king (321–c. 297 BCE), may have met Alexander; certainly his armies moved into territory held by Alexander's forces after the Macedonian leader's death.[11] By the time of his own death, Maurya's empire covered the whole of northern India from sea to sea. It reached its height under his grandson Ashoka, who controlled almost the entire Indian subcontinent, including most of the mountainous regions of North India. Only the kingdoms of Chola, Chera, and Pandana, on the tip of the Indian subcontinent—what is now more or less Kerala and Tamil Nadu—withstood Maurya forces, and even they paid monetary tribute to the empire.

These victories were won in bloody battles in which thousands were killed. But unlike nearly all other rulers throughout history who became more bloodthirsty as they went from victory to victory, Ashoka was overcome with shame and remorse when he realized the extent of the killing: "a hundred and fifty thousand people were deported, a hundred thousand were killed and as many as that perished," he lamented.[12]

In his chagrin over what he had wrought, he looked for a different path, and came upon that of the Buddha. Convinced of its righteousness, he sent delegations of Buddhist missionaries around the known world, preaching harmonious morality. It was in this period that Buddhism first came to China, to Southeast Asia, and into the high mountains of Eurasia, to the west and north. Ashoka ordered rock monuments and pillars erected to encourage his subjects to be more generous, kind, and moral, and to inform them of the measures he had taken to reform society. More than thirty of these monuments survive. Most are in northern India, although one still stands in the South. Another in Afghanistan, written in Aramaic and Greek, indicates how far and wide the power of the Mauryas extended.

But the South resisted. How much this independence was due to crafty statesmanship on the part of southern leaders, how much to their military force, and how much to the sheer luck of being out of the main lines of conflict, will probably never be known.

Geography surely contributed. The forces of the Maurya Empire, as well as those who later carried the banners of the Gupta dynasty, of Islam, and of Moghul power, began their conquests in the North on the plains of great rivers, the Ganges and the Indus, which have their origins in the Himalayas. Farther south, however, the Ghats and and the Nilgiri Hills impeded conquerors because their large cavalry forces could not easily enter the southernmost part of the country.

Only one South Indian power, the Pallava dynasty, whose influence extended from 400 to 1200 CE, appears to have roots in the North, and even it seems to have arrived in the South not as an invading force, but by making love, not war—that is, through marriages with local royal families.[13]

At the time when the Roman Empire faltered, the Cheras were locked in a struggle with two other would-be South Indian empires, the Cholas and the

Pandyas. The former had their initial seat of power along the Kaveri River itself. They had expansionist ambitions nearly as big as the North Indian powers, and set out warships and troops during their more than four-hundred-year rule. Chola monuments and records show that for a time they ruled part of the eastern Indian coast as far as the Ganges, and also conquered Sri Lanka, the Andaman and Nicobar Islands in the Indian Ocean, and parts of Southeast Asia. The temples and statuary they left behind demonstrate how sophisticated their culture was.[14] Their rivals, the Pandyas and the Pallavas, at various times also controlled places far from South India.

In addition, South Indian culture and religion was carried by myriad unnamed merchants from the South, the largest part of the formidable wave of Indian traders who influenced other societies around the Indian Ocean and beyond. Hindu monuments show up on the Indonesian island of Java dating from the third century CE, even before the main thrust of the Chola and Pallava expansionary projects. And the Vietnamese, you'll remember, were stalled for centuries in the Red River Valley because of the Champa Empire, which at the beginning was resolutely Hindu and whose writing used a script that was related to those used in South Asia, not China.

Yet, perhaps ironically, it was through the port cities from which Hindu and Indian civilization expanded around the Indian Ocean that the last conquest of India began. This time the South was the entry point for emissaries of the European powers, and mountains and high plains proved to be no defence.

As noted before, pepper had been well known and coveted in Europe for centuries, but there were other riches sought after by what we now consider "the West." The Periplus of the Erythraean Sea, a document written around the time of Strabo to guide traders and navigators around the Indian Ocean, speaks of the jewels and gold that could be had on the Indian coast.[15] Getting these trade goods back to Europe involved a sea voyage, followed by an overland trip across the Arabian Peninsula or into Egypt. The desire to find a more direct route that could be controlled by Europeans lay behind the voyages of exploration sent out by European kings beginning in the fourteenth century.[16]

The Spaniards backed Christopher Columbus, who sailed west. The Portuguese financed several other navigators who went east. One of them was Vasco da Gama, the first to make the long trip around Africa to India. He brought back a cargo of riches shortly after Columbus returned from the New World with a rather scanty haul. Da Gama made three trips to the East Indies at the beginning of the 1500s. He died on the last one, in 1524, and he was buried in a church in Fort Kochi, in Kerala.

Back up a bit: before I went to India I was confused by Fort Kochi, Kochi, and the difference between Cochin and Cochinchina, the name the French gave to the southern part of what's now Vietnam. At first glance there appeared to be no connection at all; and yet there is a real one that I began to appreciate only after I paid a visit to one of Kochi's most famous tourist attractions. I came away with an even greater appreciation of how ideas are transmitted as people move around the world.

Fort Kochi sits on the south shore of the estuary through which the River Periyar flows into the Arabian Sea. Before the independence of India in 1947

The Indian state of Kerala has a network of lakes, lagoons, backwaters, and canals (like this one in Fort Kochi) so beautiful and productive that it has been a stopping place for traders and adventurers for centuries. *Photo provided by author.*

it had passed through Portuguese, Dutch, and British control over a period of about five hundred years. But, if you watch fishermen raising or lowering big fishing nets into the tides at the mouth of the estuary, you'll be told that long before Europeans a fleet arrived from China with much treasure, including these nets.

The Malayalam name for the nets—*cheena vala*—tells a different story, which underscores the twists and turns of cultural influence that affect the daily life of ordinary folk even as kingdoms and governments seek riches and power. The term appears to be a direct borrowing from the Portuguese, in which *China* has an *i* that is pronounced *eee*, and *vala* means sail. Fishermen in Vietnam use almost exactly the same technique to fish in the waters of the Mekong Delta, and it's likely that the nets were imported from there, but by Portuguese emissaries, not Chinese ones.

What is more—and this is the source of my confusion about names—the Portuguese were responsible for calling the southern part of the Southeast Asian peninsula Cochinchina. When they arrived off the coast of what's now

The Chinese Nets are a big tourist attraction at Fort Kochi in Kerala, but they were brought to India by the Portuguese, it seems. *Photo provided by author.*

Vietnam, they had recently begun trading in Cochin (or Kochi, which is the current Indian spelling.). As they travelled farther east, they heard the very similar sounding Malay term for the Indochinese peninsula that in turn was derived from what the Chinese were then calling Vietnam: Gia Chi (or Jiaozhi in Mandarin Chinese). Arab traders had transformed that into Kawci or Kawci min Con, meaning "near China." The Portuguese heard "Cochinchina" and thought the name was meant to distinguish the Southeast Asian peninsula from the Cochin they knew on the western coast of India.[17]

Whatever, after I had watched the nets for a while I continued to explore Fort Kochi on foot. It was hot: old trees planted along the quiet streets filtered the sun, but did little to buffer the heat. Then I came upon a stucco-covered church that wouldn't be out of place anywhere that emissaries from the Iberian Peninsula landed. A tour bus was just leaving so I felt no reluctance about escaping the sultry afternoon inside, where I found a plaque commemorating Vasco da Gama, who had been buried in the church.[18]

At first I was a little puzzled by its relatively austere decoration: wooden pews, several ceiling fans, and in front an altarpiece made of dark, carved wood. No stations of the cross, no crucifix, so I knew that I wasn't in a Roman Catholic church. Rather strange, I thought: if the Portuguese brought Christianity to these coasts, their legacy should be the Church of Rome, shouldn't it?

What I didn't realize was that the church, now dedicated to Saint Francis and part of the Church of South India in the Anglican Communion, had gone through many changes since it was built in 1503, shortly after da Gama's second voyage—and that it was far from being the first Christian church in India. Rebuilt several times, it became a Protestant church in 1663 when the Dutch took Kochi from the Portuguese, and in 1825, after the British succeeded the Dutch, it became part of the Church of England.

But when the Portuguese arrived, there had been Christians in southern India for nearly fifteen hundred years, a fact that has repercussions for why Kerala and Tamil Nadu are different even though they share much of the same history. Saint Thomas—the disciple who at first doubted the resurrection of Christ—arrived about twenty years after Jesus's death to spread the Gospel. Roman trade in spices was at its peak, and trips from the Middle East and Europe would have been "easier, more frequent and probably cheaper than in any time" until Vasco da Gama discovered the sea route.[19] Thomas apparently converted local Brahmins to the Good News and built seven churches in settlements north of Kochi. Ever the missionary, though, after he planted

Christianity there he travelled across the Western Ghats into what is now Tamil Nadu, where his welcome was dramatically less warm. He was opposed there by Brahmin leaders and, so the story goes, was killed by a Hindu when he prayed at the base of a cross erected on the top of a hill not far from the coast.

Fly into Chennai today (formerly called Madras), the capital of Tamil Nadu, and, just to the south of the international airport, you'll see the complex of church buildings sheltering that very cross, which is now an official national shrine. The green hilltop stands out, surrounded as it is by the crowds and bustle of Chennai, the fourth-largest city in India (population 7.088 million, according to the 2011 census). The current chapel was originally built by the Portuguese in 1547, has been refurbished several times, and now is home to Roman Catholic worshipers. The sanctuary has a curved roof—"shell patterned," says the official description, forming "a tunnel of spiritual energy leading thousands of pilgrims in fervent prayer to the cave of their hearts."[20] Saint Thomas himself is supposed to have been buried nearer the coast and not far from the Adyar River: the San Thome Basilica, a Gothic-inspired structure finished in 1896, stands over his tomb.

Note that I said Saint Thomas is "supposed" to have been buried there. Few records dating back that far still exist, because the Portuguese destroyed everything about India's early Christians that they could find, even though they also found a large stone cross that in the sixteenth century supposedly bled on feast days. They considered the Christianity practised in India to be heresy, even though references in works found elsewhere make it clear that before Christian communities were alive and well in South India long before the Iberian Peninsula became Christian. As for Saint Thomas making his way to India: that is considerably more plausible, given the Roman spice trade, than the story of the beheaded body of Saint James sailing around the Iberian Peninsula to Compostela in a stone boat that is part of Santiago de Compostela lore.[21]

Today, Christians make up 18.38 percent of Kerala's population (with 54.73 percent Hindu and 26.56 percent Muslim).[22] Besides Roman Catholics and Protestants, this includes Christians belonging to branches linked to the Syrian Church that were apparently begun when another Thomas, he of Cana, arrived in the fourth century CE.[23] Until seventy years ago Kerala also had a small but ancient Jewish community, which had arrived in two waves. The first came after the destruction of the Second Temple in Jerusalem in 72 CE, about twenty years after Saint Thomas arrived in Kerala. The second came with the Portuguese, or somewhat later, when the official zealotry of the Inquisition saw Jews chased from the Iberian Peninsula. The Pardesi Synagogue, not far from the St. Francis Church in Kochi, was built by them. Still in use, it often has trouble attracting enough worshippers to hold services, since most of the local Jews immigrated to Israel when the new Jewish state was formed a year after India gained independence.[24] Note, too, that what appears to be the first mosque built in India was constructed in Kerala in 629 CE—three years before the Prophet's death and ten years before Islam arrived in Egypt—at the behest of the Keralite king who had embraced the new faith.[25]

The religious mixture in Tamil Nadu is and has been considerably different. The overwhelming majority—87.8 percent—are Hindu, while 5.86 percent are Muslim and 6.12 percent are Christian. That may be due to the strong traditions of past local empires like the Pallavas and the Cholas, who resisted "foreign" religions, including Islam, more vigorously. On the east coast, the Portuguese faced resistance from organized Muslim fisherfolk, with the result that their beachhead was more precarious than on the west coast. In addition, the prevailing winds and sea routes made it more tempting to continue eastward across the Bay of Bengal toward Southeast Asia and China than it was to try to subdue the Tamil-speaking people in what is now Tamil Nadu. Nevertheless, about the same time that the Jesuit Alexandre de Rhodes was producing a Portuguese-Latin-Vietnamese dictionary using a Romanized alphabet, another Jesuit was working on translating sacred texts into Tamil.[26] Several Portuguese words also show up in modern Tamil, but the European impact on southeastern India appears to have been much less than on the west side until the British entered the picture in the eighteenth century.

The British East Indies Company, that quasi-governmental body that governed much of India for so long—just as the Hudson's Bay Company controlled a great swath of Canada—had its first Indian toehold at Machilipatnam, where the River Krishna enters the Bay of Bengal after travelling from its headwaters in the Western Ghats. The settlement's name comes from words for "fish" and "settlement," suggesting that it was a hub for fishing early in its history. By the time of the Greeks, though, fine cloth was being produced nearby. The Hellenes called it muslin, after their name for the town, Maisolis. Twelve centuries later, Marco Polo also praised the cloth originating there: it was, he said, "of the finest texture and highest value....There is no king or queen in the world who would not gladly wear a fabric of such delicacy and beauty."[27]

Because textiles were one of the things the English wanted to trade, it's not surprising they aimed to set up shop where the fine cloth was readily available. Soon, however, they found themselves "cramped and worried by the Mohammedan authorities," as James Talboys Wheeler wrote in 1886.[28] The

British wanted to build a fortified settlement "and mount it with British cannon without the interference of local authorities." Therefore they negotiated a lease on a narrow strip of land six miles long and one mile deep, situated about three hundred miles south of Machilipatnam, a spot which happened to be not far from the site where Saint Thomas was martyred. The parcel included an island just off the coast, where the British built Fort St. George. On the mainland, Talboys Wheeler writes, "weavers, washers, painters, and hosts of other Hindu artisans, flocked to the spot and eagerly entered the service of the British, and began to set up their looms and to weave, wash, and paint their cotton goods in the open air beneath the trees. Villages of little huts of mud and bamboo soon grew up on the sandy soil to the north of the island and factory." The whole settlement came to be known as Madras, and for nearly two hundred years much of South India was administered from there as the Madras Presidency.

Textiles are still important to the economy in Tamil Nadu, but it also has become a hub for heavy manufacturing, and Chennai is sometimes called the Detroit of India because of its auto industry. The city nevertheless has invested heavily in a new metro system, which opened in 2015 and whose

The British ruled most of southern India from the Madras Presidency, whose administrative centre was at Fort St. George (seen here in 1858) in what is now Tamil Nadu. *Source: Wikimedia Commons, public domain.*

trains whiz along thirty-five kilometres of track. Kochi, with an urban population of only about a quarter of that of Chennai according to the 2011 census, also has recently opened a metro.

These two modern transport projects are evidence that South India as a whole is far from backward. Comparative statistics for literacy, infant mortality, and the ratio of women to men show both states to be among the most advanced in the country. According to the 2011 census Kerala has a literacy rate of 93.91 percent, and Tamil Nadu, 80.33—far above the country average of 74 percent.[29]

Both states also boast great improvements in infant mortality rates, a basic indicator of good health care. The rate was 48 deaths per 1,000 live births in Tamil Nadu in 1998–9, but dropped to 17 per 1,000 in 2016, half the national average of 34. Things were even better in Kerala: 10 per 1,000, which can be compared to rates of 5.8 in the United States and 4.6 in Canada.[30]

Another instructive population statistic is the ratio of men to women. Barring selective abortion or female infanticide, about 97 girls are born for every 100 boys. The ratio is nearer to 100:100 by about age five because boys seem to be more fragile in early life. Then, in societies where fewer women die in childbirth, the percentage of females pulls ahead that of males among adults. Therefore the total male-to-female sex ratio in a population is often used as a rough measure of both health care and the place of women in it: the more women to men, the healthier and more egalitarian it is. By this measure, Kerala does better than Tamil Nadu: the 2011 census showed 108.4 females per 100 males in Kerala, but only 96.6 females to 100 males in Tamil Nadu.[31] That figure is considerably better than the ratio of 94.3 females to 100 males in the Indian population as a whole,[32] but the difference raises the question: Why?

The answer probably lies in part in the fact that large segments of society in pre-colonial Kerala were matrilineal—inheritance came through one's mother—which made girls more prized than they were elsewhere in India. On the other hand, in Tamil Nadu Hindu and Muslim laws dealing with inheritance prevailed for centuries, which has had implications for the value placed on women in society.[33] As noted in the previous chapter on Tunisia and Algeria, while the inheritance rights given women under Quranic law were progressive at the time they were introduced, they are now considered unfair by many. And while the inheritance rights of Hindu women improved drastically with a major reform in Indian family law in 2005,[34] women still face discrimination in Hinduism. A protest by some five million women in

Kerala kicked off 2019 when two women were barred from entering a Hindu temple, despite a Supreme Court ruling that they had the right to do so.[35]

Nevertheless, progress in both South Indian neighbouring states has been so impressive that political commentators have mused about what would happen if the South, seventy years after independence, were to rally around their common linguistic heritage and secede from India to form an independent Dravida Nadu. The idea isn't new: during the run-up to Indian independence in 1947, the points of correspondence and the common history shared by South Indians was enough for a vocal group to propose making a distinct country of the parts of India where Dravidian-derived languages are spoken. It would have encompassed what is now Kerala and Tamil Nadu, as well as the states of Karnataka, Andhra Pradesh, and Telangana.

In the end the idea was nixed, but, according to Palash Ghosh of the *International Business Times*, were an independent polity made up of Kerala, Tamil Nadu, and the two other states that speak Dravidian languages to be formed, it "would have a population of about 250 million, less than the United States, but greater than Russia, Brazil or Pakistan." When Ghosh wrote these words in 2013, the four states contributed "22 percent of India's national GDP and...28 percent of national employment. The region has a GDP of about $300 billion US (about the same as Southeast Asian powerhouse Malaysia). Moreover, South India's GDP is projected to...edge near $650 billion by 2020 (larger than the current economic strength of Switzerland and just below that of Saudi Arabia)."[36]

(A note here is necessary about that other political entity where Tamil is spoken, the island country of Sri Lanka. Decades of civil war between the Sinhalese majority and the Tamil minority ended in 2009. Today, the Sri Lankan government appears uninterested in closer relations with Tamil Nadu: conflicts over fishing grounds have seen Sri Lankan navy boats harassing fishermen from Tamil Nadu in the strait separating India from the island.[37])

Seventy years after Indian independence the Dravida Nadu separatist cause remains a force in Tamil Nadu's politics, and love of the Tamil language, spoken by eighty million people—more than Italian, Thai, or Polish—is impressive. As Whitney Cox wrote in the *New York Review of Books*, this attachment to the language "shades over into an intense loyalty, even a quasi-religious devotion, of a sort rarely seen elsewhere."[38] The separatist Dravida Munnetra Kazhagam won the state elections in 1967 and since then, parties advocating "greater functional autonomy" within the current Indian constitutional framework have dominated the state's politics. In 2016 the All India Anna

Dravid Munnetra Kazhagam won re-election, in part by drawing on continued support for "national affirmation," as the idea of a special place for those who speak a Dravidian language is now called, but also by making handouts to voters directly, which is a feature of Tamil Nadu's politics.[39]

In Kerala, on the other hand, that kind of vote buying is not practised, nor is the idea of "national affirmation" a factor in the political discourse. What marks Keralan politics is the translation of left-wing economic and political theory into myriad education and health programs. In 1957, the state became the first political entity anywhere to democratically elect a Marxist government, doing so more than a decade before Chileans voted in Salvador Allende in 1970. Left-wing coalitions have alternated since then, with more centrist alliances involving the Congress Party, the winner in 2016 being the Communist-led Left Front, which trounced a coalition headed by Congress.

One of the fruits of this emphasis on social reform is the most educated population in the country. An indicator, perhaps, is what I saw when I was there: along the street leading to the train station in Kochi, vendors of peanuts and candies squatted on the sidewalk reading local newspapers while they waited for business. This is the only developing country I'd been in where the street vendors pass their time reading rather than in conversation. Kerala's language, Malayalam, is spoken by only 32 million people, less than half of Tamil Nadu's population of 72 million (2011 census), yet two of the ten most-read newspapers in the country (which in 2018 was estimated to have a population of 1.3 billion) are published in Malayalam, and 60 percent of Keralites read newspapers regularly, compared to 16.5 percent nationally.[40]

Kerala is cited as an example of what can happen when a society invests heavily in education, health, and equality, although whether the model can be adopted and adapted elsewhere has been hotly debated for a half-century, as has whether doing that, were it to succeed, would be a good idea.[41] The state has India's third-lowest poverty rate, but also, perhaps counterintuitively, the third-highest unemployment rate.

"We have poverty, but not abject poverty," Dr. Rajan Chedambath, director of the Centre for Studies in Culture and Heritage of Cochin, told me when I was there in 2005.[42] Many skilled and educated Keralites work in Saudi Arabia and other countries of the Middle East because they can make much more money there than at home. What they send back, in fact, makes an enormous difference in the local economy, many observers believe. Not only can families buy more consumer goods with money earned offshore, but the amounts they are willing to pay for land has driven up the price of urban and rural lots substantially.

Very little rice is now grown in Kerala because the former rice fields can be sold for higher prices to Keralites who've worked abroad than if the land were sold for agricultural purposes. Nor do many Keralites stand in line for low-paying jobs at home. Instead, these jobs are filled by a steady influx of people from other Indian states, including Tamil Nadu, who are ready to work at menial tasks for wages that are higher than those in their home states, but not high enough to attract Keralites.[43]

After a time in Kochi and its environs, I began to think I could tell at a glance the difference between Keralites and outsiders, who tended to be shorter, thinner, and even darker: the ones camped alongside the railroad track, the women drying saris on low bushes on a Sunday afternoon, the fellow who insisted on carrying my suitcase at the train station, and also, I'm told by a municipal official who practically whispered the news, the ones responsible for what crime there is in Kochi.

That last statement raised my hackles: typical xenophobic prejudice, I thought; probably no basis in fact. The kind of thing you run into everywhere because people frequently want to bolster their own position by looking down on neighbours, a factor in politics that we'll run across more than once as we consider neighbouring states. Burundi and Rwanda are a worst-case scenario, while a milder version will raise its head in the chapter on the United States and Canada. But prejudice against outsiders exists, and suffice to say that in the Indian context, it means that the cause of a united Dravida Nadu was lost a long time ago, if it ever had a chance.

Listening to the rhetoric, I was reminded of arguments about the place of French and English in Canada, which, as we'll see, has had profound effects on the country. Nationalism can take on layers and layers of additional meaning as politicians use it to further their own agendas.

BRAZIL AND **THE REST OF SOUTH AMERICA**

The language we were urged to learn growing up in California was Spanish. Tijuana was just twenty miles away from where we lived in San Diego, and the state's history was full of references to Spanish and Mexican missions and ranches. In sixth grade we had weekly Spanish lessons, and in high school Spanish was the usual choice of a foreign language for kids who were headed to university.[1] But since then my Spanish has been buried under the languages I've tried to learn more recently: German, French, and, for the last ten years, Portuguese. That's why I was surprised at how easily Spanish surfaced in 2013 when I took a bus trip from Cuzco, Peru, to Rio Branco, Brazil.

The purpose of the trip was to look at the way a new transcontinental highway linking the Atlantic and Pacific was changing things, as well as at traffic and public transit in the Brazilian cities of Brasília and Curítiba.[2] Because I'd been concentrating on my struggle to learn Portuguese for so long—it's easy to read for anyone who knows another Romance language well, but oh, the differences in pronunciation!—I hadn't really thought about dusting off my Spanish. Then I found myself on a bus, pulling out of Cuzco with a young man next to me and twelve hours of travel in front of us before we reached the other side of the Andes.

He was curious about me—people usually wonder what a grandmotherly *gringa* is doing travelling alone—and so started the conversation. To my astonishment I understood him. It took some rooting around in the lower levels of my brain, but after a few minutes we were conversing. Not perfectly, to be sure, but well enough to learn quite a lot about each other.

That night in Puerto Maldonaldo, Peru, not far from the Madre de Deus River in the Amazon Basin, I understood what the hotel staff was saying, although I must admit I was relieved to speak English when the front desk recognized my accent and switched to it.

Otherwise it was Spanish all the way until, on the second day of the trip, we neared Brazil. Portuguese began to appear then, and at the international boundary we were searched by two teams of border patrols. The Spanish-speaking Peruvian one physically opened every piece of our luggage to search for—I guess—contraband. Fifty metres away, the Portuguese-speaking Brazilian team loaded each piece on a conveyor to be scanned by a portable X-ray system. The two methods took about the same time and I wondered if there were some sort of message hidden there, if not in our suitcases. Something to do with technology, and equipment, and, perhaps, the need to be up to date when, like Brazil, you're the biggest kid on the block.[3]

The bus trip began to feel much too long when night fell on the second day. The steward—this was a high-class bus—kept a steady stream of Hollywood action movies playing, sometimes dubbed in Portuguese, sometimes in Spanish. As I struggled to understand the scratchy soundtracks, I began to think about the reasons lying behind the seemingly arbitrary dividing line between the two countries and the two languages. The border between Brazil and Peru doesn't follow the height of the land, as does most of the boundary between Kerala and Tamil Nadu. Nor was one single river the dividing line, the way the Bến Hải was in the era of the two Vietnams.[4] Determining the boundary between Peru and Brazil was obviously more complicated: this stretch of the border wasn't fixed until the mid-twentieth century after much discussion and a couple of border skirmishes, I learned later.[5]

— — —

Setting the boundary between what would become Peru and Brazil started out to be a simple thing. All it took was a pope half a world away to divide the planet into Spanish and Portuguese realms by fiat. His action is a brilliant example of how political expediency thousands of kilometres distant can seal the fate of nations.

The official portraits show Pope Alexander VI as a worldly (and portly) Spaniard, with a huge nose, swarthy skin, and a penchant for gold and jewels. Born at a time when Europeans were looking for new ways to reach the riches

Pope Alexander VI divided the world between Spanish and Portuguese realms, thus setting the stage for the division of South America into Spanish-speaking and Portuguese-speaking regions. *Source: Bode-Museum, Berlin, Germany. Artist: Unknown 15th-century artist. Creative Commons.*

of the Orient, he became a cardinal at the age of twenty-five, and went on to make a name for himself because of his shrewdness in business, his learning, and the fortune he amassed. Then, just as Christopher Columbus was leaving on his first voyage to the Western Hemisphere, he became pope.

By that time, as we've seen, Portugal had a head start in the grand exploration boom that sent ships from Europe to look for new ways to bring home the riches of the East. Pioneering the *volta do mar*—sailing west to pick up favourable winds in the South Atlantic to carry them around Africa—the Portuguese were already bringing treasure back from India by sea within a decade of Columbus's return.[6] Their jump on the Spanish was a result of the fact that, long before Spain became Spain, the Portuguese were able to drive the Moors from their part of the Iberian Peninsula and consolidate an independent kingdom.

To understand how that happened, some backstory is necessary. Iberia had long been on the maps of early civilizations. The Phoenicians traded there. Like Carthage on the North African coast, Cartagena, the port in modern Spain, owes its name to the Phoenician term for "new city," *qart-hadasht*. Later, the Romans ruled the entire Iberian Peninsula for hundreds of years. But after Rome fell, in the fourth century, the next couple centuries of disorder left the door open to invasion by the forces of Islam, who, as noted in the chapter on Tunisia and Algeria, advanced with lightning rapidity around the Mediterranean.[7] The Prophet died in 632, and seventy-five years later North Africa had been conquered; the Moorish advance into Europe was only stopped definitively in 732 at the Battle of Tours, just over the Pyrenees, in what is now France.

Local rebellions occurred almost immediately, but the *Reconquista*—the winning back of Iberia by the region's Christian inhabitants—proceeded by fits and starts.[8] The extreme west, Galicia and Portugal, were the first to throw off Moorish rule. In 1139 the Kingdom of Portugal was proclaimed, and a hundred years later the last of the Moors were expelled from a Portugal whose frontiers have changed very little since.[9]

In contrast, Spain took much longer to become completely independent of Moorish power and to present a united face to the world. It took a royal marriage—that of Isabella of Castile and Ferdinand of Aragon in 1474—for most of the rest of the peninsula to be brought under one rule. Eighteen years later, when Columbus set sail toward the West, the last of the Muslims—and Jews—were banished.

Pope Alexander VI had been instrumental in creating Spain back when he was Rodrigo Borgia (yes, one of *those* Borgias). It was he who masterminded the union of Isabella and Ferdinand and their dominions. She was the half-sister of the king of Castile, who for reasons of his own wanted to marry her off to the much older king of Portugal. But in addition to being old, the Portuguese king was also ugly, and no one should have been surprised when she balked. She was only sixteen, after all, with an abundance of curly blond hair, fine features, and a confident glance that shows up in her portraits all her life, even after her hair darkened and her body thickened. When messengers brought back the latest news about young Prince Ferdinand of Aragon, who already was the king of Sardinia and in line for more titles, she fixed her heart on him.[10]

But the course of true love never did run smooth, as Shakespeare would say somewhat later.[11] The young lovers were second cousins, and so forbidden to marry without special dispensation from the sitting pope. Because the couple's advisers wanted a quick marriage for political reasons there wasn't time to get the proper documents from Rome. Enter Rodrigo Borgia to the rescue: he produced a papal dispensation dated a few years earlier that would allow Ferdinand to marry a cousin. There appears to be no doubt that the document was a forgery, but then Borgia was not one to play by the rules. The fact that Rodrigo sired four children whom he recognized as his after he took his priestly oath of celibacy—among them Cesare Borgia, who Niccolò Machiavelli considered an exemplary prince in his manual for gaining and keeping power, *The Prince*—is only one indication of his disregard for the usual rules of moral conduct. (He had other, unlegitimized children, and several mistresses too, but that's not part of this story.)

Isabella, however, was kept in the dark about the forgery because Borgia and the others knew that she did take her religion seriously and that the devout princess might back out of the marriage even though she was deeply enamoured of Ferdinand. The pair, who became known as the Catholic Monarchs, welded the various parts of Spain into a nation, and looked kindly on the proposal of a Genoese navigator who'd sailed extensively with the Portuguese.

Christopher Columbus had tried first to get the Portuguese king to bankroll his expedition to the west, which he thought would lead him to the Orient. His timing was bad, though, because just when he was preparing to make his pitch in Lisbon, the Portuguese navigator Bartolomeu Dias arrived home to announce that he'd made it around the tip of Africa, proving it was possible to get to India by going south and then east. Hearing that exciting news, the Portuguese king saw no reason to finance a chancy western expedition, since going the other way was now a proven success.

Of course, it turned out that success could be had by going west too. Columbus made landfall on the island of Hispaniola (about which we'll have more to say) in October of 1492, and subsequently claimed all the land he came upon for the Spanish Crown.

Needless to say, when the Portuguese learned of this they were not pleased, particularly since more than ten years before a papal decree had given all the lands south of the Canary Islands, off the coast of Africa, to Portugal. The Portuguese king dashed off a stinging protest to Pope Alexander VI, then newly elected, who replied with two more papal bulls. In essence, the first gave the Spanish everything to the west of an imaginary line in the Atlantic a hundred leagues west and south of any of the islands of the Azores or Cape Verde. A few months later the second bull gave to Spain all territory, mainland and islands "at one time or even yet belonged to India," even if east of that previously demarcated line.[12]

The Portuguese were even less pleased.

It should be remembered that when all of this was happening very little was known about what land might be waiting just over the horizon, and no one really knew how to reckon just how far east or west a ship had sailed. (The problem of longitude wouldn't be solved for another three hundred years, in fact.) What is more, neither the Spanish nor the Portuguese, and certainly not Pope Alexander VI, had any idea that the riches of the East were more than thirteen thousand miles away by the westward route, and that two great continents stretching north to south and nearly from pole to pole barred the way. Yet what was clear was that the Portuguese, with their growing fleet of seagoing vessels and their knowledge of the winds and currents of the oceans, would be more than a match for the Spanish, should the two countries come to blows over sea routes.

In the end, representatives of the two countries began negotiating on their own. The Portuguese agreed in principle to the idea of a dividing line, but they insisted on pushing it considerably westward. In the end the line was set 370 leagues west of the Cape Verde islands. (Since a league is generally considered to be about 3 miles, the total distance would be about 810 miles or 1,303 kilometres.) The Treaty of Tordesillas (named after the place where it was negotiated) was ratified by both sides, and then confirmed by another papal bull in 1506.[13]

By that time Alexander VI and Isabella had died, and the consequences of the shifting of the line were becoming apparent. In 1500 Portuguese ships under the command of Pedro Álvares Cabral were blown off course as they manoeuvred south in the Atlantic in order to round Africa. Cabral was surprised to see a mountain looming to the west and made for it. After determining to his satisfaction that the land lay east of the famous line, he claimed it for Portugal, and sent one of his ships back home with the news announcing the "discovery."

It's been suggested that the Portuguese king had reports from earlier mariners that a large body of land lay to the west, in the southern Atlantic, which made him so insistent on moving the line of demarcation westward. Others suggest that Spanish mariners had landed on the great eastern hump of South America a little earlier. Whatever: the outcome was that the agreement about the division of the world was enough for the Portuguese to claim a huge chunk of South America—and to insist, for a while at least, that Newfoundland and Labrador were also Portuguese territory.[14] That last claim did not last very long—too cold, some Portuguese venturers are reported to have said. Nevertheless the Portuguese legacy is reflected in place names on the island of Newfoundland like Fogo (Fire) Island and two Portugal Coves, not to mention the name of the mainland portion of the province, Labrador— named, it appears, in honour of João Fernandes, a small landowner (*lavrador*) and explorer from an island in the Azores who was given a patent from the Portuguese Crown to explore and settle the land in 1499.[15]

— — —

The Portuguese maritime presence continued late into the twentieth century because of the cod fishery off the Grand Banks of Newfoundland. Portuguese sailing ships known as the White Fleet set off from Lisbon and the Azores to fish until the mid-1970s, as they had for five hundred years, but at that point huge factory ships from other countries began scooping up cod, and very soon the fishery collapsed.

One of the last of these sailing ships was in St. John's Harbour in 1971 when my husband and I visited there. He was there for an academic conference at Memorial University, and I was along for the ride. The ship had crossed the Atlantic under sail, its decks covered with small dories from which the men jigged for cod in the cold ocean waters. But at that point—this was early June—the sails were furled and the men were on shore for rest and to resupply.

It took us a while to figure out just what they were doing there. We were young whippersnappers with a high opinion of our intellect, but none of us had seen anything like it. As we wandered around the dock area, a few fishermen started a pick-up soccer game in the parking lot. Four or five of the academic types invited themselves in, using pantomime. The Portuguese graciously allowed them to play, but they were much better than the Canadians. At the end there were handshakes all around, but few words exchanged, because nobody could understand the fishermen. No one was even sure what language they were speaking.

And therein, quite probably, lies the explanation of how the Portuguese were able to maintain their identity in the New World, far beyond the wildest dreams—or nightmares—of Alexander vi or Ferdinand and Isabella. Portuguese is not Spanish, despite its similarities, and is in fact an older language.

A word about languages in general, however. Why do people speak different languages, anyway? All of us are descended from a very small population of early anatomically modern humans who must have spoken a mutually intelligible language. Study of the world's languages today indicates that the same concepts show up again and again, and that words conveying their meanings are frequently linguistically related. One list of such words gives a snapshot of what early humans must have been concerned with. The top ten entries are: I, you (singular), this, who, what, one, two, fish, dog, and louse![16] Today, however, there are about sixty-five hundred languages around the world, with at least twenty-five hundred in Africa, where our ancestors all started out.[17]

Be that as it may, it's clear that when people live separately, the way they communicate with each other changes slowly but surely over time. Just how and why this happens is the subject of much debate, but it's clear that how people speak is extremely important to what they think about themselves. How else did the Vietnamese continue to be Vietnamese when faced with the force of Chinese culture and power? Same thing in India: the Dravidian languages of the South reflect the unique history of societies stalwartly different from those of North India, and also serve as the link between the two unidentical twins of Kerala and Tamil Nadu. And I'm convinced that a major reason why Canada and the United States remain distinct polities is that at the beginning of British North America Francophones made up about 30 percent of the population, and that, even though the percentage has dropped, it is still more than one in five.[18] While many in the rest of Canada would abhor being incorporated into the United States, the hard fact of such a sizeable non-English-speaking population has stymied American imperialism over two hundred and fifty years of history. But more about that later.

Both Portuguese and Spanish developed from a mixture of Latin spoken at the end of the Roman Empire on the Iberian Peninsula, remnants of the languages spoken by the region's pre-Roman population, and invaders from the north who swept in after the fall of Rome. Both languages also carry traces of the Arabic that came when Muslims took most of the peninsula.

At the time that Pope Alexander vi set about dividing the world, Portuguese was a much more established and standardized language. It—and not Latin— was declared the language of study in 1290, when King Deniz created the first Portuguese university, located first at Lisbon before it was moved to Coimbra. However, it wasn't until the mid-fifteenth century that a grammar of Spanish was even written. Queen Isabella reportedly didn't see much point in it when she was presented with a copy.[19]

But one papal bull five hundred years ago was not enough to determine what language people would be speaking in the new-found land of the Western Hemisphere once it was settled. Another treaty set boundaries on the other side of the world, giving the Philippine islands to Spain. Later, when the enormous size of South America became known, the two colonial powers worked out agreements that gave most of the watershed of the Amazon River to Portugal and settled border issues between what would become Brazil and its Spanish-speaking neighbours to the south, Uruguay, Paraguay, and Argentina.

This despite the fact that for a period of sixty years—from 1580 to 1640— Portugal was ruled by Spain. Philip ii of Spain, the grandson of Ferdinand and Isabella, turned out to be the only surviving heir to the Portuguese Crown after a couple of plot twists that can only happen when you rely on passing down leadership through blood relations. To make a long story short, Sebastião, the Portuguese king from 1557 to 1578, led a disastrous invasion into North Africa and disappeared, apparently killed (although for decades his followers persisted in believing that he might come back). The next in line was a Cardinal who, unlike Rodrigo Borgia, took his vows of celibacy seriously and had no children. When he died, Philip, whose mother was a Portuguese princess, inherited the throne. Luckily for the fate of the Portuguese language, Philip spoke it well, and the arrangement allowed a certain amount of independence for Portugal. Yet it took a small war before Portugal became completely free of Spain.

By that time, Spanish explorers were well launched on their conquest of the Indigenous peoples of Central and South America. Hernán Cortez crossed Mexico, leaving from the Caribbean coast, and then fell upon the riches of the Aztec Empire. Francisco Pizarro also started from the Caribbean, and subsequently crossed the northern part of South America, to strike southward. He and his men were travelling in pursuit of the gold and silver rumoured to be found in the Inca Empire, which stretched from what is now Venezuela to the present-day Chile.

When it came to setting up permanent beachheads, the Spanish were most successful around the Caribbean and down the Pacific coast south of the Isthmus of Panama. Veracruz, on the east coast of Mexico, was founded in 1519, as was Panama City on the Pacific coast. There followed settlements in Venezuela, Colombia, Ecuador, and Peru during the 1520s. The Portuguese, of course, had dibs on the eastern edge of the continent, but they were slower to start building cities. The first ones in Brazil date no earlier than 1533, largely because so much Portuguese energy was initially directed eastward toward India and the Far East.

The neat division of the world between Spain and Portugal would not last, of course. The age of exploration was also one of increasing religious ferment, with the authority of the church questioned not only in newly Protestant countries like England and Holland, but also in France. Just because a pope had divided the world didn't mean that Protestant nations were going to abide by his decisions. Nor was the French king about to let the riches of the world slip away. Not long after Cabral reported the discovery of Brazil, French vessels were exploring its Atlantic coast and the establishment of a colony was attempted between 1556 and 1567.

Today the only remnant of these other European powers' interest in South America is the triumvirate of small states on the Caribbean coast: the former British colony of Guyana, the former Dutch colony of Suriname, and Guyane (French Guiana), which is still a part—a *collectivité territoriale*, somewhat like a state in the United States or a province in Canada—of France.[20]

On the other hand, Spanish-speaking South America is now divided into fifteen countries, reflecting in large part the administrative divisions that the Spanish established early in their empire. Lima, Peru, became the centre for trade, law, and power, and also home to the first university in South America, the National University of San Marcos, established in 1551. In the next hundred years several other universities were established in what are now Argentina, Bolivia, Colombia, Chile, and Ecuador, making it possible to educate an elite without young men having to take the arduous journey back to the mother country for training. In contrast, Brazil, despite a population that grew faster than the Spanish-speaking one on the other side of the continent, remained much more closely tied to its mother country. A sea voyage from San Salvador or Rio to Lisbon took at most three months, which meant that communications were much more efficient, so many services could be directed from Portugal.[21] There is no more telling evidence of this than the fact that no university was established in Brazil before the beginning of the twentieth century.[22] Some doctors and military engineers were trained at schools set up in Bahia and Rio in the early nineteenth century, but for all intents and purposes men who were headed for professional careers had to go to Portugal. This usually meant the Universidade de Coimbra, where, you'll remember, King Deniz set the tone for Portugal's development as a polity separate from the rest of the Iberian Peninsula when he declared that the language of instruction would be Portuguese.

That might have been enough to assure Portuguese as the language of Brazil, but another powerful European sealed the continent's linguistic destiny: Napoléon Bonaparte, the little Corsican soldier who became emperor of the French from 1804 until 1814, and then briefly in 1815. His intention was to conquer the world, and for a time he certainly had much of continental Europe under his control. He set up members of his family as rulers in his stead, among them his brother Joseph, with whom he quarrelled, but whom he nevertheless appointed king of Spain in 1808.

Reaction in Spanish-speaking colonies was complicated. Many *criollos*, as the descendants of the Spanish born in the New World were called, were upset at the usurpation of the Spanish throne by a French puppet, but at the same time they saw the turmoil as the moment to press their desire for independence. They'd been figuring out how to run their own affairs for more than three hundred years, and they had the examples of the United States of America and Haiti to the north as new nations that had cast off colonial rulers.

Between 1810, when Chile founded its legislative assembly independent from Spanish rule, and 1825, when Bolivia declared its independence, nine sovereign countries were formed in South America. The new nations' boundaries for the most part followed the administrative, judicial, and military lines created decades previously by the Spanish to govern their colonies.

The men who led these wars of independence were sometimes larger than life. One of the chief architects of Latin American independence, Simón Bolívar, is a case in point.[23] The son of a very wealthy Spanish family that had settled in what is now Venezuela in the sixteenth century, he was orphaned at nine, and was strongly influenced by his guardian, who was a partisan of Enlightenment ideals. This was reinforced by travels in France and England, where Bolívar read Hobbes and Hume, Montesquieu and Rousseau.

When he returned he was swept up in the agitation against Spanish rule. The Guerra de la Independencia Española began in 1808 when Napoléon sought to conquer the Iberian Peninsula, installing his brother Joseph as king of Spain. The descendants of the Spanish in Latin America seized the opportunity to fight for a measure of independence from the mother country.

Bolívar had dreamed of making one country of Spanish South America, and to this end he led forces into battle over a period of twenty years. But he died bitterly disappointed that squabbles and local power struggles doomed what he called the Federation of the Andes. A united Spanish-speaking counterweight to what would become the colossus of Brazil never developed. Nonetheless, he gave his name to one of the countries formed—Bolivia—as well as the currency of Venezuela, the *bolívar*, and innumerable parks, squares, and monuments throughout the region.

The story in Brazil was quite different. The American Revolution was the inspiration for an early independence movement whose leader, Joaquim José da Silva Xavier (called Tiradentes), was hanged and quartered in 1792. But rather than result in more revolutionary activity, as it had in Spanish South America, the rise of Napoléon tied Brazil more closely to Portugal. The advance of Napoléon's forces on the Iberian Peninsula led the Portuguese royal family to flee at the end of 1807, just before Lisbon was taken by the French. Escorted by British naval vessels, the royal flotilla took fifty-four days to cross the Atlantic carrying with it as many as fifteen thousand courtiers and servants, as well as much treasure. British forces finally pushed the French from Lisbon after five years of war, but the king and court tarried in Brazil until 1821, when a new constitution was approved by the Portuguese Cortes. There followed some bloody conflicts between factions in the royal family over who would rule in Brazil and who in Portugal. In the end Dom Pedro II became emperor of Brazil, ruling until 1889, when a republic was declared.[24]

— — —

Another time, another pope. Three million people at Copacabana beach! Singing priests! Nuns wading in the surf! Young people from all over the world keeping an all-night vigil, fueled not by alcohol or drugs but by religious enthusiasm! Then in the morning, as grey clouds hung over the mountains that frame Rio de Janeiro, the pilgrims to the twenty-eighth World Youth Day brought sunshine by cheering and singing and praying at a mass led by Pope Francis. The newly elected pontiff was causing a sensation.[25]

I wasn't there: my visit to Brazil came four months later. But the country was still surfing a glorious wave then. Brazil would be home to the World Cup of soccer the following year, and Rio the Olympics two years after that.

Few suspected that political scandal and economic calamity loomed just over the horizon, so euphoric was the mood. The visit of the first pope from the Americas was eliciting enthusiasm orders of magnitude larger than the usual excitement of a papal visit.

This was the first pontifical visit of Pope Francis anywhere. He'd only been elected the previous March but was well on his way to putting his own stamp on the office. He is apparently a man who believes in living simply and was called "the world's parish priest" before he was elevated to the Roman Catholic Church's highest office. All this contributed to the overwhelming welcome. What's more, his home country of Argentina is just south of Brazil, and quite clearly many of the people who greeted him considered him "one of ours" as he rode around in an open popemobile, waving, shaking hands, and kissing babies.

Pope Francis is a man cut from cloth quite different from the ermine-trimmed velvet robes worn by Pope Alexander VI. Both the church and the world have changed a lot since Alexander and his fiats divided the world between Spain and Portugal. Yet even with this new, modest, twenty-first-century

Pope Francis's first official visit anywhere was to World Youth Day in Brazil. The trip culminated in a mass on Rio's Copacabana beach on July 23, 2013. That he is more at ease in Spanish than Portuguese speaks volumes about linguistic divisions in South America. *Source: Agência Brasil. Creative Commons Attribution 3.0 Brazil licence.*

pontiff, the shadow cast by the crafty old scion of the Borgia dynasty still loomed over the South American landscape. The division between Spanish and Portuguese realms remains clear hundreds of years after Alexander VI proclaimed his papal bulls.

Nowhere was that more evident than in the addresses Pope Francis gave during his time in Brazil. Videos show him rather stiffly reading messages in Portuguese: his official biographies say he can speak it—it's near the end of the list of the languages he speaks—but he looks and sounds uncomfortable. Then there is the half-hour interview he gave to Brazil's major media chain, Globo.[26] In it, a much more relaxed pontiff jokes with the Brazilian journalist that, even though there is rivalry in South America, everyone gets along. After all, he says, the pope is Argentinian and God is Brazilian, making reference to a hit Brazilian comedy in which God decides to take a holiday and comes to Brazil to find a vacation replacement.

It's a small, irreverent joke, and certainly not the kind of discourse you'd hear from Alexander VI, were he to be transported to our times. More importantly, note that Francis delivers it in Spanish and speaks that language throughout the interview, even though he clearly understands the questions in Portuguese. The difference between the languages is so clear-cut that the Brazilian broadcaster felt compelled to subtitle the interview in Portuguese when it was aired.

So the division of the world continues in a very important way. The most intimate part of the lives of hundreds of millions of people—the language in which they love and dream and think and, yes, pray—is what it is because of what a pope decreed five hundred years ago.

HAITI AND **THE DOMINICAN REPUBLIC**

But let us return once more to the fifteenth century, when Rodrigo Borgia was the newly elected Pope Alexander VI and Christopher Columbus had just landed on the Caribbean island of Hispaniola. That island, now shared by Haiti and the Dominican Republic, was "a marvel," the great navigator thought. He wrote back to the Spanish authorities who had bankrolled his expedition:

> There are marvelous pine groves and broad meadows, and there is honey and there are many different kinds of birds and many varieties of fruit....The sierras and the mountains and the plains and the fields and the land are so beautiful and rich for planting and sowing, for raising all kinds of cattle, for building towns and villages....[The trees] were as green and as beautiful as they are in Spain in May, and some were in flower and some in fruit, and some at another stage according to their nature, and there where I travelled the nightingale and other birds of a thousand kinds were singing in November.[1]

Sounds like heaven on earth, which did not surprise Columbus because, along with riches, he had expected to find Eden. "The sacred theologians and learned philosophers well said that the earthly Paradise is at the end of the Orient," he wrote home, and so that's what he saw in the verdant islands of the Caribbean.[2]

Take a trip to Hispaniola today, or look at satellite photographs, and you'll see how much things have changed. From high above, the cities and roads built by the people of the two frenemy nations (who now number nearly twenty million) are clearly visible, but, even more strikingly, you can see that the forest has nearly all disappeared on the Haitian side of the border. It is as striking a boundary as there can be.

— — —

I had a glimpse of that change in 1973 when I tagged along when my sister and her husband went to Haiti. It was the first foreign country that I ever visited, aside from a few hours spent in Tijuana, Mexico, just across the border from San Diego where I grew up. Had my sister not asked if I wanted to accompany her while her husband scouted some resort locations for his employer, I would never have thought of going. But why not? It was February, and Montreal in February can be very cold—certainly not paradise.

The flowers of Port-au-Prince—the bougainvillea, the lantana, the geraniums, the calla lilies, the plumbago—were indeed lovely. Their colours and their scents took me back to my childhood years in Southern California, making me almost homesick for a time that ordinarily I am glad I left behind. Had I realized that all were imports, gifts from the botanical explorers that followed in Columbus's wake, I might have felt even more ambivalent about them.

But, no, I didn't understand what they meant, nor did I appreciate the burden the women coming down from the hills carried on their heads. They were walking barefoot along the roads to town when I went out shortly after sunrise. Most balanced baskets of charcoal on their heads, although some carried other things too—watercress, for one. But the charcoal they were carrying to market was essential for the functioning of the city since it was used for cooking everywhere by ordinary people.

No hint of that in the nice hotel where we were staying, of course: the excellent meals were cooked on propane. But we did get a glimpse of the impact charcoal harvesting was making on the environment when we took a couple of excursions. From a hilltop lookout we could see how many of the surrounding hills were skinned back, with low bushes the only vegetation. Then on the way to a beach resort we passed over a plain where the road was washed out in places; in others it was covered with a slick of mud that had flowed down from bare hills in the distance.

Several decades later the area of deforestation has grown exponentially and now runs all the way up to the Dominican border, in the mountains at the centre of the island. In Haiti only 2 percent of the original forest is left, while in the Dominican Republic a significant portion of that forest—or healthy second-growth forest—remains. The graphic divide says much about the difference between the twins and their uneasy relationship since discovery by Europeans.

As recently as a century ago the situation was much different. Then an observer who came to study the forest resources noted that the "least known of the Greater Antilles, Columbus's island of Hispaniola, has been the least changed since pre-Columbian times. At least 75 per cent of the land area is still forest-clad with trees that can be classed as timber."[3] The author noted that much was being cleared for small-plot agriculture, but had great hopes that the forest could be successfully managed.

It wasn't to be, particularly on the Haitian side. The reasons for this go back to that voyage Columbus took, which changed so much of the world.

Satellite photo of the border between Haiti and the Dominican Republic, showing the difference in forest cover between the two countries. *Source: NASA, public domain.*

— — —

The land was not empty when the Spanish ships arrived. The Taino people had been there for centuries, even millennia. Manioc (also called cassava), a plant known to have been domesticated in South America, was a staple in their diet, suggesting—as does analysis of their language—that they reached the Caribbean from the south. Some of the words used by them, such as *barbacoa* (barbecue), *hamaca* (hammock), *kanoa* (canoe), *tabaco* (tobacco), *batata* (sweet potato), and *juracán* (hurricane), are ones that have been borrowed by languages around the world. *Ayti* was their word for their island. It means "land of high mountains," of which there are several: one, Pico Duarte, is more than three thousand metres (or ten thousand feet) high. These linguistic relics, alas, are among the few things still extant of the Taino culture, since their population crashed within a few decades of first contact with Europeans. Like most other peoples in the Western Hemisphere, they had little or no immunity to diseases the conquistadors brought with them. The first recorded outbreak of small pox in the New World occurred in December 1518 on Hispaniola, and the disease was carried in the following months by Spanish forces to Mexico, where it killed thousands.[4] The Indigenous peoples on the island were also worked pitilessly by the Spaniards, who began requiring tribute on a regular basis within a few years of their arrival.

As the Taino died, the Spaniards had to look farther afield to get workers for the sugar cane and other plantations that they established. The first shipload of African slaves arrived at the beginning of the 1500s, not much more than a decade after Columbus's landfall. The stage was set for a colony in which a large non-European population was in bondage, while the Europeans were the landowners, the managers, the literal slave-drivers—and mostly male. The first crops were sugar cane, tobacco, and indigo, plus foodstuffs to be shipped to other Spanish beachheads that had not had time to establish their own farms. Another "product" was mixed-race children: since very few European women came in the early years, the Spanish grandees had liaisons with Indigenous and African women.

Then the wave of Spanish exploration and exploitation moved west to Mexico and south to the western coast of South America. Hispaniola became less important to the Spanish. The only settlement that remained was Santo Domingo, on the southeast corner of the island. Founded in 1496, it is the oldest continuously inhabited European city in the Western Hemisphere, but not long after its founding it became merely a stopping place for ships headed in and out of the Caribbean. (Centuries later it would become the capital of the Dominican Republic, but that is getting ahead of the story.)

The western end of the island was to all intents and purposes abandoned by the end of the sixteenth century.[5] Animals left behind—cattle, goats, sheep, and horses—had free run of the valleys and hills. As time passed, pirates or privateers—particularly French ones—who preyed on shipping in the Caribbean, appropriated a smaller island off Hispaniola's north coast to use as a base; from there they made forays to the big island to hunt the animals gone wild. They cooked what they caught over an open flame—*boucaner* in French, related to the modern Québécois term for smoke, *boucane*—and, voila!, they became known as *boucanniers* or buccaneers. In time, they also set up more permanent settlements on Hispaniola, which set the stage for a piece of diplomatic gamesmanship that would determine the island's subsequent history.

By the middle of the 1600s the camps of the buccaneers had become more permanent settlements. Augmented by adventurers who'd decided not to go back to France, the French presence grew, particularly after the establishment of the French West India Company in 1664.[6] This annoyed the Spanish, who nevertheless did little about it. But in 1697 the Spanish wanted to negotiate a treaty to end a war with the French, English, and Dutch that barely touched the Caribbean. In a tit for tat, they gave up one-third of Hispaniola in exchange for, as they say in sports trades, other considerations. (The war was called King William's War in English-speaking parts, but more about that when we turn to the unidentical twins of Ireland and Scotland. In North America, the borders set by the treaty would also have great implications, but that will have to wait too.)

Under the French, the colony prospered—or at least it made many of the French masters rich. To find labourers for its sugar-, cotton-, and tobacco-producing plantation economy, the French imported an estimated one million slaves from Africa, many of whom died because of terrible conditions. Nevertheless, by 1791 Haiti had a population of about 500,000 slaves, of whom probably 300,000 had been born in Africa, and 50,000 free people, of whom perhaps were 30,000 were black or mixed race. This was in sharp contrast to the population in the Spanish part of the island, which stood at about 125,000 in 1790, with 40,000 white landowners, 25,000 black or mulatto freedmen, and 60,000 slaves. The difference in population mix is largely due to the fact that the economy in the Spanish-held territory was based on agriculture for local consumption, so far fewer slaves were imported to work plantations.[7]

Note that year: 1791. France was two years into its revolution, with its Legislative Assembly sitting for the first time, and agitation was growing to get rid of the king. In the newly created United States of America, George Washington entered the third year of his mandate as the first president, and Vermont was accepted as the fourteenth state in the Union. In Great Britain, Thomas Paine, whose *Common Sense* was a battle cry during the American Revolution, published the first volume of *The Rights of Man*, saying that popular political revolution is permissible when a government does not safeguard the natural rights of its people. (Simón Bolívar took great encouragement from Paine in his campaign to liberate Spanish-speaking South America by the way.)[8]

It also was the year that dissatisfaction with the status quo in Haiti boiled over into open rebellion. Events and ideas from elsewhere probably contributed, but rage over the lack of rights and the institution of slavery itself had been simmering for a long time. Communities of runaway slaves had developed in the mountains over the years, and skirmishes between them and landowners were not uncommon. While the latter were far better armed, many of the runaways were seasoned warriors, skilled fighters in Africa who had been unlucky enough to be captured in local wars. The African victors enslaved them, and eventually sold them to European slave traders.

The rebellion in 1791 began somewhat differently from these skirmishes, however, in that free people of colour—not runaway slaves—were the major instigators. They wanted more rights and freedoms, and some historians say that initially the fight was between the whites and the free blacks. But the mass, the slave population, joined a few months later. White colonialists were well-prepared and well-armed, but the first round of fighting saw four thousand whites and perhaps a hundred thousand slaves dead in the fray.

The uprising, while it threatened France's source of sugar and its place in the Caribbean, was viewed with considerable favour by France's own revolutionaries, and by 1794 slavery was outlawed in all French territories. And yet peace was not gained. The fight continued for another ten years. In the meantime, after Napoléon Bonaparte took the reins of government, France went back on its word regarding the abolition of slavery, and French troops were dispatched to Haiti to put down the rebellion. In the background the intrigue among European powers continued, and in 1795 Spain ceded its portion of Hispaniola to France under the Treaty of Basle as part of another settlement of a generalized conflict involving Great Britain, France, and Spain.

Finally in 1804 Haiti—with almost the same boundaries it has today—declared its independence from France. It was the first nation in the Western Hemisphere after the United States to become independent from a European power and the first black free nation. Tens of thousands of former slaves had been killed by then, and most of the whites who survived the conflict had left the island quickly, so the population was much reduced.

But peace did not come. Because the new country's leaders feared the return of French forces, the Haitian military was expanded and then maintained: up to 10 percent of able-bodied men were henceforth in active service, which was a great drain on any attempts to rebuild the country. In addition, after 1825 Haiti had to pay reparations to French slaveholders. The initial amount was set at 150 million francs (variously estimated at between $660 million and $1 billion in modern Canadian dollars), which was subsequently reduced to 60 million (or about $264 million).[9] The claim bankrupted the Haitian treasury, and, somewhat ironically, the new country was forced to take out a loan from French banks. Historian Laurent Dubois says that by 1898 "fully half of Haiti's government budget went to paying France and the French banks. By 1914 that proportion had climbed to 80 percent."[10]

On the other side of the island, a movement for independence developed in the wake of Napoléon's march through Europe. At the same time that the Portuguese king and his courtiers were leaving for Brazil in the winter of 1807–8 one step ahead of the French army, Spanish loyalists seized the opportunity to wrest the eastern end of Hispaniola from France with the aid of British forces, who were also fighting Napoléon. The colony once again became Spanish, and remained so until, inspired by Simón Bolívar and his allies fighting for independence in South America, a group of dissidents declared an independent state of Haiti Español in 1821. That came to naught in the short term, however, because the Haitians invaded and occupied it for the next twenty-two years. The period proved crucial for future relations between Haiti and the Dominican Republic. In a study done for the US State Department, Richard A. Haggerty notes:

The occupying Haitian forces lived off the land in Santo Domingo, commandeering or confiscating what they needed to perform their duties or to fill their stomachs. Dominicans saw this as tribute demanded by petty conquerors, or as simple theft. Racial animosities also affected attitudes on both sides; black Haitian troops reacted with reflexive resentment against lighter-skinned Dominicans, while Dominicans came to associate the Haitians' dark skin with the oppression and the abuses of occupation.[11]

So the end of the nineteenth century saw two nations occupying Hispaniola that hated each other, both with few exports and with economies based mostly on subsistence farming. A small, educated, usually light-skinned elite governed Haiti and would have liked to reintroduce plantation agriculture in a big way. But in what might be seen as a revenge of the people, a pattern of *lakou* developed. (The term is Haitian Creole—or Kreyòl—and is related to the French *la cour*, or yard.) These were small homesteads where an extended family could grow their own food as well as a small cash crop like coffee or sugar cane to buy what couldn't be produced at home. The plots of land had been acquired after the end of slavery, and, according to Dubois, for generations the homesteads were successful in providing for the great majority of Haitians. But over time the holdings were divided into smaller and smaller parcels, and the fertility of the land declined through overuse. The fertility of the people continued unabated, however, and so the population grew.

Small farms were also the pattern in the Dominican Republic, but since the country started with a much smaller population and also was somewhat larger in area, population density was lower. Consequently the toll taken on the land by intensive subsistence farming was less severe.

Both countries were viewed as pawns by the United States, who wanted to make sure that the Caribbean was an "American lake," particularly once construction of the Panama Canal—opened in 1914—began. Both were occupied by the United States in the early twentieth century, the Dominican Republic between 1916 and 1924 and Haiti between 1915 and 1934. US control continued long after, however: an arrangement begun in 1905, through which the United States collected customs revenues for the Dominican Republic in order to ensure that debts to United States interests were paid off, lasted until 1940.[12] While some infrastructure was improved during this period, the local population was less than happy. The American occupiers with rare exceptions spoke only English and showed little sensitivity to local customs or—and this is particularly important in Haiti—the complex relations of class and colour.

The end of the occupation did not mean good times, however, and certainly not good relations between Haiti and the Dominican Republic. In the former, political instability and lack of a policy directed toward either economic development or sustainable agriculture led to a continued reliance on small farms, which provided the bare minimum to the rural population. In addition, factional conflicts resulted in witch hunts for enemies.

Things were not rosy on the Dominican side of the border either, but many Haitians decided the slightly better conditions were worth crossing over into it, usually illegally. This did not please Dominicans, and growing tensions climaxed in 1937 in what came to be known as the Parsley Massacre. The attack on Haitians was encouraged, if not orchestrated, by the Dominican dictator Rafael Trujillo. He is recorded as saying just before the violence that he had been investigating complaints of cattle thefts and raiding by Haitians along the frontier (which had only been definitively surveyed in 1929). To those who were therefore "prevented from enjoying in peace the products of their labor, I have responded, 'I will fix this.' And we have already begun to remedy the situation. Three hundred Haitians are now dead in Bánica. This remedy will continue."[13]

As we shall see in the next chapter, on Rwanda and Burundi, official sanction of attacks on a perceived enemy is a powerful tool. The result this time was as many as ten thousand Haitians killed, with many more pushed back to the other side. In biblical fashion, victims were selected according to whether they could pronounce a shibboleth, in this case the Spanish word for parsley, *perejil*, in the Dominican way—Creole and French speakers couldn't get either the *r* or the *j* right.[14]

Later in the twentieth century, one of the greatest politicians of the democratic left in the Dominican Republic was the son of Haitian illegals who had earlier fled over the border. Adopted by Dominican peasants, José Francisco Peña Gómez became mayor of Santo Domingo, and ran three times for the presidency. Today the airport in Santo Domingo is named after him.[15]

— — —

But the Parsley Massacre lives on in the current troubled relations between Haiti and the Dominican Republic. In 2013 Dominican courts retroactively denationalized Dominicans of Haitian descent, extending all the way back to 1929.[16] The blow was somewhat softened the following year by a law allowing people of Haitian descent to "re-register" if they could prove that their parents had been in the Dominican Republic legally. According to Amnesty International, by mid-September 2014 nearly 300,000 people had applied, but shortly thereafter forced deportations began. By May 26, 2016, 44,000 people had been deported, and another 66,000 had returned "spontaneously."[17] Then, between January and July 2018 alone, 80,832 Haitians were either deported or turned back at the border.[18]

Some blame the current anti-Haitian attitude on racist sentiments rampant in Dominican society: despite the mixed-race heritage of most Dominicans, black faces have always been near the bottom of the pecking order. As Dominican-American writer Junot Díaz says: Dominicans tend to have "sensationally racist generalizations" about Haitians.[19]

Among Haitians, the Parsley Massacre is a rage-producing memory that has had long-lasting effects. "We have voices sealed inside our heads, voices that with each passing day grow even louder than the clamor of the world outside," writes Haitian-American writer Edwidge Danticat in her novel *The Farming of Bones*.[20] This critically acclaimed work tells the story of a young Haitian woman living on the Dominican side of the border whose lover is killed during the troubles.

The instigator of the massacre, Rafael Trujillo, was a ruthless man who killed his enemies and who saw enemies everywhere. He revelled in self-aggrandizement. He had the name of the capital changed from Santo Domingo to Ciudad Trujillo, and that of the highest mountain on Hispaniola from Pico Duarte to Pico Trujillo. Churches were ordered to display the slogan "Dios en cielo, Trujillo en tierra," or "God in Heaven, Trujillo on Earth"—and eventually a decree was issued to reverse the order so as to put Trujillo first. He had friends in high places internationally, and tens of millions of dollars in foreign bank accounts. He also took a cut from the wages of civil servants, and even the earnings of prostitutes. The result was a subjugated society, in which everyone, it seemed, worked for the dictator. As Nobel Prize winner Mario Vargas Llosa writes in *The Feast of the Goat*, a novel about the life and death of Trujillo: "In this country, in one way or another, everyone had been, was, or would be part of the regime....The worst thing that can happen to a Dominican is to be intelligent or competent...because sooner or later Trujillo will call upon him to serve the regime, or his person, and when he calls, one is not permitted to say no....The Goat had taken from people the sacred attribute given to them by God: their free will."[21]

For a while the United States looked kindly on his regime, since it seemed staunchly anti-Communist at a time when American foreign policy was determined by Cold War calculations. But as Canadian diplomat John W. Graham, who was posted to the Dominican Republic at the time, reports, "the entire country was quivering with fear" and trouble was just beneath the surface.[22] In the end, Trujillo's scheming and megalomania came back to haunt him: in 1961 he was assassinated by the very men who were supposed to protect him.

In Haiti, François Duvalier, a medical doctor who became known as Papa Doc, took power not long before the end of Trujillo's rule. He had been a leader in the fight against infectious diseases like typhus, yaws, and malaria in rural areas after studying public health at the University of Michigan in the 1930s. He served as minister of health in one government, but left it to go back to medicine following a coup d'état in 1950. Strongly influenced by the idea of *Négritude*—the intellectual movement begun by black writers in Paris in the 1920s that glorified the talents and stories of Africans in Africa and in the diaspora—he opposed the government of the day while in hiding, and then announced his candidacy for president in 1956 before it was ousted. When it fell, and presidential elections were held, Duvalier won 679,884 votes to his opponent's 266,992; and, while there were allegations of fraud, the election was one of the fairest in the nation's history.[23]

An advocate of black populism, Duvalier deposed many highly placed mulattos in the civil service and military, actions much appreciated by the vast majority of the population, who resented the way that light-skinned, mixed-race people had controlled the country for generations. Among his partisans was the father of Quebec writer and member of the Académie française Dany Laferrière. A very young history professor, the elder Laferrière was appointed mayor of Haiti's capital, Port-au-Prince, in 1957, and then served as secretary of commerce and industry. When he declared that shops hoarding food in protest against the Duvalier regime could in good conscience be looted, he was sent abroad as Haitian ambassador to Italy and then Argentina. But shortly afterwards Duvalier began his purge of anyone who might cross him, and the strong-minded Laferrière was high on Papa Doc's list. He went into exile in New York, never returning to Haiti before his death in 1984.[24] After his flight, reprisals against him were perceived as such a threat that young Dany, then age four, was sent out of harm's way to live in the country with his grandmother. He never saw his father again.

Duvalier grew more and more erratic and vengeful as time went on. His apologists say that he appeared to have undergone neurological damage after a nearly fatal medical incident in 1959 that caused a personality change. A diabetic, Duvalier suffered a heart attack that was diagnosed first as a diabetic emergency. He was given insulin but ended up in a coma for nine hours. Recovery took weeks, and during his recuperation the reins of power were taken by the leader of his web of spies and henchmen, the Tonton Macoute, the quasi-military force whose name comes from a Kreyól term for the boogeyman, the uncle—*tonton*—who carries a big bag—*macoute*—in which he steals you away. When Duvalier recovered, he suspected his second-in-command of plotting the whole thing, and had him killed. Then, because there were rumours that the man had transformed himself into a black dog, Duvalier ordered all the black dogs in Haiti killed.

What weirdness! the reader might well say. Voodoo, derived from the animist religions of Africa, has long been practised in Haiti but also to a lesser extent in the Dominican Republic. (Note that it is a living force as well in Louisiana, Jamaica, and Brazil.) One of the reasons given for the greater presence of African culture in Haiti is that the particular intensity and cruelty of the Haitian plantation system killed so many slaves within a few years of their arrival. This meant constant importation of new slaves from West Africa, who brought with them fresher memories, more immediate African experience, and a closer connection to African belief systems. It's no coincidence that the Voodoo pantheon resembles that of the pantheistic religions practised on what was once called the African Slave Coast: Kreyòl grammar also has much in common with languages spoken in West Africa.[25] In contrast, Christian clergy had more success in pushing aside African beliefs in other areas of the Caribbean, like the eastern end of Hispaniola, where plantations were less common and fewer slaves were imported. Furthermore, after the Haitian Revolution the Roman Catholic Church refused to recognize the new nation until 1860, which meant that no priests worked in Haiti for nearly fifty years, leaving a spiritual vacuum. Therefore it's not surprising that Voodoo's rich mixture of magic and the sacred became such a factor in Haitian life.

(As for those who might scoff at Voodoo's legitimacy: remember that, even today, in a country as progressive as Iceland, roads are rerouted to avoid places where elves live, that ancestor worship is taken extremely seriously in Japan, and that some American football players make a fetish out of praying for divine help before games.[26])

— — —

The year my sister, her husband, and I went to Haiti Papa Doc Duvalier had been dead for three years, and his son—not quite twenty-three—was in charge. Exactly what that meant, I had no idea, but it seemed that there was a general optimism among the people we met. It was a non-representative group, to be sure: cab drivers and waiters, a few locals in the bar, women I spoke to in the market early in the morning, workers attached to international charities, a journalist or two. But Haiti was poor—very, very poor—and my sister, who had travelled quite a bit in Europe but never in a Third World country, was overwhelmed. We had to stay a week because of our airplane tickets and flight schedules, but she insisted her husband not leave us alone in Port-au-Prince. We could enjoy the pool and the rum punches, as well as a few well-organized day trips, but she didn't want to venture out of the hotel unless we were well protected.

And I must admit that I, too, would have been flummoxed by what I saw had I not recently read *A House for Mr. Biswas* by V.S. Naipaul. His novel was set in Trinidad, but the feel of the place was similar: the heat, the plants, the small houses in poor repair, the woman lying beside the road one afternoon with the baby wrapped up next to her, her hand stretched out, begging. Now I know that Trinidad's history is quite different from Haiti's, but at the time Naipaul's book gave me a dim idea of what was going on. So I decided to explore, to not waste this glimpse of a place that was so unlike what I was used to. As soon as the church bells rang at about sunrise, I went out by myself. Nobody would bother me early in the morning, I figured. The bad guys would be sleeping it off, the solid citizens would be on their way to school or work. It is a strategy I've used again and again as I've travelled, and in a curious way I'm grateful that neither my sister nor her husband felt up to joining my exploratory walks on that first trip beyond my North American comfort zone. Otherwise I wouldn't have discovered how much one can learn from them.

On that trip I talked to a young woman on her way to an office job, to schoolgirls headed for class, to a young man in a white shirt and tie who asked me if he could help me find my way. With my fractured, rudimentary French— we'd only been in Montreal a few years at that point, and I still had a long way to go in mastering that language—I could understand much of what they said, and if I scrunched up my eyes and tried to read them aloud, the signs in Haitian Creole made some sense. Later I realized that I was lucky that guessing was still a large part of my French-language acquisition strategy. Because I wasn't concerned with niceties of spelling or grammar, I also didn't worry that the Kreyòl spelling has little to do with French because it was standardized by Americans in the 1940s and the letters all have English values. Only later did I realize how that debases an important part of Haitian culture.

— — —

And then there were the women carrying charcoal on their heads.

Jared Diamond calls Haiti and the Dominican Republic a "natural experiment." Starting by considering today's glaring difference in vegetation on either side of the border, he explores why that developed and ultimately concludes that a major reason was the very different decisions made by the strongmen who wielded such power in each country.[27]

Since Trujillo ruled the country as if it literally belonged to him, self-interest led him to stop others from logging even though he and his cronies continued doing so. For a while after his assassination, even that restraint was lifted and the Dominican hills were extensively logged. Then in 1966 Joaquín Balaguer was elected president and, although he'd been one of Trujillo's henchmen and would later be responsible for his own share of political repression and torture, he surprised many by instituting environmental policies that have made an enormous difference in the country. First he banned all commercial logging and closed all the country's sawmills so that the rivers needed to run hydroelectric projects were not clogged by silt.[28] Professional foresters, both local and from abroad, were hired, and the anti-logging ban was enforced by the armed forces for a time. Balaguer was out of office from 1978 to 1986, but returned to, among other things, enact a suite of programs that got people to shift away from the use of charcoal for cooking and heating toward propane. There's an irony here, of course. The world is currently going through the motions of weaning itself from fossil fuels in order to reverse the increase in greenhouse gases, but in the Dominican Republic switching to a fuel derived from hydrocarbons made a huge impact on the health of the country's forests and the environment in general.

Given Haiti's disorganization, measures like Balaguer's would have been very hard to put into effect, even if its governments had wanted to, and certainly the Duvaliers, *père et fils*, did not. Nor have things subsequently improved on that front. Since the end of the Duvalier days—Papa Doc's son, Baby Doc, was forced to flee in 1986—governments have come and gone with amazing frequency, so that planning for the long term would have been impossible, had anyone tried. The one election that probably was fairly honest, when former priest Jean-Patrice Aristide was elected, ended in a debacle when Aristide was pushed out of office by an American-backed coup.

Natural disasters have also contributed to chaos in Haiti: Hurricane Ike in 2008, Hurricane Matthew in 2016, not to mention the earthquake in January 2010, which devastated a good part of the country. Recovery has been hindered because of considerable political unrest. The country was without a president for more than a year after an election in 2015 was declared void because of electoral fraud. In a new election, held in November 2016, Jovenel Moïse was elected with 55 percent of the vote, but the results weren't formally confirmed for nearly a year, well after he had been sworn in as president.[29]

In contrast, government in the Dominican Republic, while sometimes corrupt and inefficient, has been much more stable, with sitting president Danilo Medina handily winning a second term in 2016.[30]

Both countries have also experienced great out-migration in the last fifty years. Today, somewhere around 1.1 million immigrants from the Dominican Republic and 700,000 from Haiti are living in the United States.[31] In the 2016 Canadian census, 165,095 people identified themselves as Haitian-Canadians, but only 23,130 said they were Dominican-Canadians.[32] The effects of this diaspora have been felt differently in the two countries because the people who left, particularly in the 1960s and '70s, were quite different. Jared Diamond mentions the great baseball players who have come from the Dominican Republic, and certainly there is no equivalent from Haiti. More significantly, the Haitian diaspora initially included far more professionals than did the Dominican one. By the 1970s, 80 percent of Haiti's professionals were in exile; indeed, in 1967 there already were 131 Haitian doctors in Montreal, and more Haitian nurses and psychiatrists in that Canadian city than in Port-au-Prince.[33]

To some extent this early wave of Haitian immigrants found Canada a much more congenial destination than the United States since the health-care and education systems in Canada were expanding at that time. Haitian professionals all spoke French—the mass of Haitians might speak only Haitian Creole, but any Haitian with any education at all learned French along the way. Therefore they found it much easier in Canada, particularly in Quebec, to have their credentials recognized than did their compatriots in the United States.

For many the past weighed heavily, however. A case in point was the father of Michaëlle Jean, who would grow up to become governor general of Canada and, afterwards, secretary-general of the Organisation internationale de la Francophonie. Her parents were staunch opponents of Papa Doc, and, even though their father was principal of an elite Protestant school, Jean and her sister were educated at home because their parents did not want them to take the oath of allegiance to Duvalier required of schoolchildren. Perhaps inevitably, Jean's father was arrested and tortured by the Duvalier regime before he fled, a "broken man" according to Jean.[34] She left with her mother and sister a year or so later, in 1967.

——— ——— ———

My sister, brother-in-law, and I saw nothing of the Tonton Macoutes when we were in Haiti, of course. The only hint of Voodoo we saw was a sanitized musical show at the hotel where we were staying. When we ventured out together it was to places like Port-au-Prince's cathedral, Notre Dame de l'Assomption. The taxi driver who claimed us as his clients took us there late one morning and then handed us over to an elderly man waiting on the front steps. The building had been badly damaged in Hurricane Inez in 1966—eight years before, mind you—but it was now on its way to being restored, he proudly told us. My experience with cathedrals up to that point was rather sketchy—I'd seen Notre Dame Basilica in Montreal but not the Notre Dame in Paris or any other European glory. Nevertheless, I wasn't terribly impressed by the soaring arches and the Gothic-style windows, which were mostly without ordinary windowpanes, let alone stained glass.

But the old man walked us through the stations of the cross and showed us where the pulpit and the choir would be when the repairs were finished. "Beautiful, beautiful," he said, words that seemed to be a major part of his English vocabulary.

We nodded, although I didn't know how to react when he indicated that we could pray if we wanted. "We're not Catholic," I said in my approximation of French.

"No matter," he said. "All one God." Then he pulled out a little container that I hadn't noticed before. "For the work here," he said. "For God."

As I remember I stepped back, naively surprised by this pitch. Until that point I'd thought his help was being given out of the sheer goodness of his heart or love for the church itself. I had no idea, no conception of the poverty he must have lived in, and my reaction is something I've regretted ever since. My brother-in-law knew better, and pulled out a few dollars as our contribution. Score one for him: he taught me a lesson.

— — —

Reconstruction of the cathedral was eventually completed and life went on. Despite the hope I had felt among the people we saw, things did not get better, though. The economic gap between Haiti and the Dominican Republic grew wider. Although before 1960 per capita income in Haiti was slightly higher than in the Dominican Republic, by 2010 the tables had drastically turned, so that per capita income was nearly ten times greater in the Dominican Republic than in Haiti.[35] Tourism, some manufacturing, coffee, and sugar were the mainstays of the Dominican economy, but in Haiti exports did not keep pace. Political conditions weren't perfect in the Dominican Republic, of course, but elections were held without massive unrest, and life was much more peaceful than in constantly disrupted Haiti.

Then in January 2010 an earthquake measuring 5.6 on the Richter scale shook Hispaniola with the epicentre in the southwest part of the island, in Haiti. The tremor was felt all over, but only the western part of the island was severely damaged. Estimates of the dead ran into the hundreds of thousands, and several million were made homeless for a period of months—or even for years. Dominicans responded quickly to the emergency, sending aid. So did countries around the world. Non-governmental organizations (NGOs) and the UN stepped in to help. Damaged houses were going to be rebuilt, and plans were made for new housing, new villages even. Trees would be planted on the barren hills; some said that the disaster could turn out to be the point at which things began to get better in Haiti.

Laurent Dubois notes in his book *Haiti: The Aftershock of History* that foreign reporters kept looking for looting, for the complete breakdown of civil society, as people were pulled from the ruins. Remarkably, there was little disorder of that sort: the level of violence was no worse than usual, he suggests. That is because the Haitian people had, on one level, set up nearly invisible (to outsiders' eyes at least) webs of trade and help that they could rely on. That was not the case for the outside efforts, he says. Outsiders generally didn't and don't understand Haiti's strengths.[36]

Solutions to Haiti's problems must come from within, Dubois insists. Homegrown talent is available, and it is ready to work even if it is in exile. An example is Georges Anglade and his wife Mireille Neptune, who left for Quebec in the 1970s, where they helped set up a new, modern provincial system of higher education while doing their own academic work. Anglade's studies of market networks in Haiti are one of the inspirations for Dubois's analysis of what already exists in Haiti. So valuable was his work that Dubois dedicates his book to him. While continuing to spend time in Quebec, Anglade and his wife were very involved in the Aristide government in the 1990s. After retirement they spent more and more time in their homeland.

— — —

The rebuilt cathedral was completely destroyed in the 2010 earthquake, as were most of the buildings in the centre of Port-au-Prince. The hotel where we stayed nearly forty years previously, the Beau Rivage, was located in that district, and I wondered how it had fared. In my research into Haiti and the

Dominican Republic, I'd come across a document that told how the building had been turned into the headquarters of the United States Agency for International Development in the 1980s. Cost overruns had been astronomical and the authors bewailed how much it had cost to retrofit the building to withstand earthquakes and bring it up to what the Americans thought were proper standards.[37] After the earthquake I spent the better part of an afternoon poring over the satellite photos on the *New York Times* website, which showed the damage. There were before and after shots of the Presidential Palace, the market, the residential areas, smashed almost beyond recognition.[38] While I couldn't determine if the Beau Rivage had been destroyed, I was overcome by photos of the interior of the cathedral we had visited so long before. The clever photographers at the *Times* had combined many photos into a panoramic montage that showed the destruction from all angles. I

The earthquake of January 2010 destroyed much of the Haitian capital of Port-au-Prince, including its cathedral. The tremor was another in a string of disasters plaguing the country. *Source: UNESCO, Creative Commons Attribution-ShareAlike 3.0 IGO licence.*

lingered over them, trying to piece together just what had stood there when the reconstructed cathedral had been a landmark, a monument to a religion that many Haitians espoused, even though they might also believe in Voodoo.

But I clicked on something inadvertently and before I knew it the photos were swirling around and around, up and down, with the sun shining most realistically in my face and the ruins looking ready to fall in on me. For a moment I was mesmerized, but then as my stomach began to lurch, I searched for a way to stop the mad progression of images, the heady, beyond-control whirl of destruction. The only way I found to get out of the loop was to close the whole thing down.

But you can't close down Haiti. It continues, as does the Dominican Republic. The struggle gets no easier, particularly as forces for change like Georges and Mireille Anglade disappear. Sadly, they both were killed in the earthquake. Then, six years later, Georges's brother Robert was shot dead while lying in bed at his home in the south of Haiti by unknown gunmen.[39] At this writing, his killer has not been brought to justice or even identified.

The shadows are long. It seems at the moment that this pair of unidentical twins is destined to grow more and more dissimilar.

BURUNDI
AND RWANDA

Early in the morning, from my hotel room in Bujumbura, the capital of Burundi, I watched women like those I'd seen in Haiti carrying loads on their heads on their way to market. My windows faced east and looked onto the hills that rise steeply from the shores of Lake Tanganyika, the second-deepest lake in the world. It lies in a trench 673 kilometres long, which is part of the great rent in the earth's surface where Africa is splitting apart. The geographic setting sometimes appeared to me to be a metaphor for what was happening seven years after the horrific genocide in Rwanda, Burundi's twin, as Hutus and Tutsis slowly inched toward a power-sharing arrangement in Burundi itself.[1]

Collecting wood for charcoal and brush to feed livestock was the cause for the desolate state of the hills before me, just as in Haiti. The Great Lakes region in the heart of Africa is even more densely populated than Haiti or the Dominican Republic, and the pressure on the land—once extremely fertile from regular infusions of volcanic ash—is immense. Indeed, overpopulation is frequently given as a major reason for the "interethnic" conflict between the majority Hutu and the minority Tutsi that has ravaged Burundi and Rwanda, the next pair of unidentical twins we'll consider.

The two countries have much in common besides a similar population mix of something like 85 percent Hutu, around 13 percent Tutsi, and 2 percent Twa, the forest dwellers sometimes called Pygmies. Today Rwanda has a slightly larger population: 12.5 million compared to 11.3 million for Burundi (2018 estimates),[2] living in about the same number of square kilometres (26,338 for Rwanda, 27,830 for Burundi.) This works out to a population density of about 402.9 persons per square kilometre in Burundi and 474.6 in Rwanda, among the highest in the world; Haiti, with its denuded, deforested hills, has 400.47 people per square kilometre.

To provide a decent standard of living with densities like that you need an agricultural economy that is efficient and supported by up-to-date technology. The Netherlands does and, even though its population density is 408 persons per square kilometre, its agricultural system is the envy of the world: in 2017 its agricultural exports hit nearly 92 billion euros, second only to those of the much, much larger United States.[3]

Burundi and Rwanda are light years away from that.

The paroxysms of violence both countries endured in the latter part of the twentieth century have many points in common. Hundreds of thousands of people—both Hutu and Tutsi—have been killed, although it's hard to tally the number. The genocidal fury in Rwanda that horrified the world in 1994 left 800,000 or more dead after three months of turmoil. In Burundi, the figure usually given for the two big outbreaks of violence in 1972 and 1993 is 200,000. Some of these figures have been substantiated by outside investigators, although they all complained about the problems of getting evidence. Suffice to say an unimaginable number of people in these two countries have been killed by their neighbours in the last sixty years. One has to look in many directions to find who is responsible for this horror. The aftermath has seen the two countries moving apart, becoming much less alike, which we'll see as we progress.

When I went to Burundi it was three weeks after 9/11, and some members of my family weren't pleased. Lord knew what might happen, given the terrorists loose in the world, they said. My sister, who had found Haiti hard to take, sent me emails ending with "this is a subliminal message: don't go." But I'd bought the ticket long before, a writing project depended on it, and besides, I'd take the advice of old Africa hands and not venture any place unless my back was well covered.

What's more, I had two things going for me: my white skin and my age. While at that point I hadn't travelled very much besides that trip to Haiti, I knew being white brought privileges most white folks aren't even aware of. Secondly, being in my late fifties was a great advantage, since I wasn't going to get hassled by young men, and in many of the places I was going, the old are treated with respect.

A third thing was at play too, although I didn't appreciate it at the time. The attack on the World Trade Center was a sophisticated, twenty-first-century phenomenon not likely to be repeated where I was going. The dangers there were much more low tech, much closer to the way people have fought each other forever. Instead of airplanes and cellphone-detonating bombs, in nearly all cases death in the conflicts of Burundi and Rwanda came in the form of neighbours killing neighbours with machetes or by burning them alive.

— — —

Bare hillsides like the ones I saw from my hotel window are nothing new in Africa. Major environmental change brought on by human activity goes back a long ways. Archeological and climatological evidence, as well as oral tradition, show that sometime about twenty-five hundred years ago, if not earlier, much of the thick tropical forests that covered the central part of African south of the Sahara Desert rather suddenly disappeared, eventually to be replaced by savannah. Not only does examination of the sediments carried into the Atlantic by the Congo River show tell-tale signs of deforestation, but excavations of a number of ancient villages show that the people living near the river were not only deeply involved in agriculture but also smelting iron for tools in a major way.[4] Indeed, by 600 BCE Central African ironworkers had developed a way of superheating air for use in a smelting process that is very much like the blast furnace technology of nineteenth-century Europe. They produced "amazingly good-quality steel—perhaps the best steel in the world of the time, and certainly equal to or even better than the steel produced in early modern Europe."[5] The fuel they used was charcoal produced from felled trees, while the land thus cleared was farmed intensively.

Some of the people who did this perfected their skill, it seems, in a corner of what is now Nigeria and Cameroon[6] before moving up the Congo River basin to the Great Lakes region and into East Africa. (They also travelled down the highlands of the West African coast as far as northern Angola, but it's their eastward movement that concerns us here.)

They spoke languages that have come to be known as Bantu. (Some linguists count up to five hundred separate tongues in that linguistic family, most of which are mutually intelligible.) They encountered people who spoke languages from another great linguistic family, the Niholitic, which includes tongues spoken on the east coast of Africa, off the Indian Ocean up through Somalia and Sudan into Ethiopia and Africa. By all accounts the culture the Bantu speakers carried with them—based on the cultivation of sorghum, millet, and yams, and, beginning early in the first millennium CE, supplemented with many sorts of bananas—was very successful. The people thrived, multiplied, and pushed ever eastwards.

Sometime near the period when the Roman Empire was waning, Bantu speakers came literally face to face with people who were not settled farmers, but who were herders like many other people in East and North Africa. In some respects it was a fortuitous meeting: herders were restricted to lands where cattle—and people—were not prey to disease spread by the tsetse fly, which thrives in the margins of forests and jungles, breeding in decaying brush and trees. As the Bantu farmed over a greater and greater area, they reduced

Cows on the shores of Lake Tanganyika. The divisions between Hutu farmers and Tutsi herders underlie ethnic violence in both Burundi and Rwanda. *Photo provided by author.*

forest cover, both to clear fields and to cut wood for fires and iron smelting. This in effect enlarged the savannah on which cattle could graze. In addition to gaining a new food source in the form of milk from their neighbours' cattle, farmers benefited from having the beasts nearby, since the manure they produced could be used to fertilize their plots.

This meeting of two cultures shows up in two surprising ways: digestion and words. Most mammals cannot digest milk after they are weaned because they no longer produce an enzyme called lactase, which breaks down a milk sugar called lactose. However, several populations of humans have specific genes that allow them to do so. They include Northern Europeans: more than 90 percent of the population of the British Isles and Scandinavia have what's called "lactase persistence."[7] But the frequency of genes carrying this trait drops where a close relationship with cows hasn't developed. It's about 50 percent in Spain, France, and in the Middle East, while among the Chinese and in parts of West Africa it is nearly non-existent.[8]

On the other hand, among Africans whose societies have included a long herding tradition, lactase persistence is extremely high. That includes both Hutus and Tutsis, even though the myth is that Hutu farmers, unlike their Tutsi counterparts, didn't have much to do with cows. But being able to digest milk gives a powerful advantage when you live near them, and the fact that most Hutus today can would suggest—even if nothing else did—that for a long time there was considerable symbiosis between the two populations.

The words people use also tell us much about their history. People don't suddenly stop speaking one language and start speaking another just because they like the sound, points out Jared Diamond in his seminal *Guns, Germs, and Steel.*[9] Communities change the language they use because they are forced to speak a conqueror's tongue, at least in official circumstances, or they choose the new one because the culture that it represents carries advantages, like new and better crops and more effective tools. We've seen an example of this with Brazil and Spanish-speaking Latin America. Several hundred languages were spoken by Indigenous people there but what you hear today in South America, at least in public, is one of the two Romance languages brought by Europeans, Spanish or Portuguese.[10] A corollary is the complex interaction of Chinese and Vietnamese: the Vietnamese used the conquerors' writing method for centuries, but clung to their own tongue, eventually pushing aside the invaders' influence. Another is the way European conquerors forced German and then French on the kingdoms that became Rwanda and Burundi—but more about that later.

The wide swath of Africa where Bantu languages are spoken is a measure of the force of the culture the Bantus carried with them. Had they been able to domesticate one or two of the large mammals they encountered as they took control of Central Africa, they might have gone farther, much farther: Diamond suggests, only half-jokingly, that a shock cavalry of Bantus on rhinos could have given the forces of the Roman Empire a run for their money. Certainly, the small North African elephants that Hannibal used to challenge the Romans as he set off from Carthage to cross the Mediterranean and then the Alps would seem to corroborate that.

But—and this is essential to understanding what has happened in Rwanda and Burundi over the last seventy-five years—Europeans for more than a century and a half looked down on the Bantus. They favoured the tall, thin, hawk-nosed Tutsi herders. One of the first Europeans to describe an encounter with them, Lieutenant John Speke of the British Indian Army, was quite taken with their appearance when he travelled into the interior from the East African coast looking for the origins of the Nile. He had no doubt that the Tutsis were inherently superior to the Hutus, postulating that the Tutsis were related to Ham and Shem, the sons of Noah, whose progeny, according to biblical tradition, went into Egypt. He thought these so-called Hamitic people were the natural elite of the Great Lakes region, and had come south and conquered the Bantus. He wrote: "It appears impossible to believe, judging from the[ir] physical appearance...that they can be of any other race than the semi-Shem-Hamitic of Ethiopia...Christians of great antiquity....Crossing the Nile close to its source, [they] discovered the rich pasture-lands...where they lost their religion, forgot their language, extracted their lower incisors like the natives...and no longer remembered the names" of the countries from which they came.[11]

Now it's quite likely that Speke was right in that these tall, filiform herders came from Somalia and Sudan, but the idea that they conquered the Bantus doesn't hold water. Firstly, recent genetic analysis shows that both Hutus and Tutsis have a remarkably similar genetic heritage, although some Tutsis do have a few genetic markers that link them to populations in the Horn of Africa.[12] But researchers today say unequivocally that had the Tutsis really been conquering invaders their subject people—the Hutus—wouldn't be speaking a Bantu language but rather one imported from Somalia or Ethiopia. That is definitely not the case. In both Rwanda and Burundi, everyone—Tutsi, Hutu, and Twa—speaks a version of Bantu, Kirundi in Burundi and Kinyarwanda in Rwanda.

I ran into the power of Bantu languages when I made a stop on my African trip at the International High Court for War Crimes in Arusha, Tanzania. During my visit I struck up separate conversations with two young men, one from Tanzania, the other from Rwanda. When the three of us met during a recess in one of the trials, I introduced them, and they tried to talk to each other; I acted as the go-between, since one spoke English, not French, and the other French, not English. Of course, I was happy to translate, but when they discovered that the Tanzanian's Swahili (another Bantu language, which also contains Arabic influences) could be understood by the Rwandan, and his Kinyarwanda by the Tanzanian, the conversation swirled beyond me until I had to ask for translation help myself.

— — —

When it comes to considering the history of Central Africa, it's worth noting that little if any Bantu influence is evident in Haiti. That is because over the centuries most of the hundreds of thousands of slaves transported to Hispaniola—as well as to the rest of the Western Hemisphere—came from West Africa, and few from Central and East Africa. A major reason for this appears to be the strong political organization that Bantu-speaking people had established in the centre of Africa. To be sure, what is called the Swahili Coast—the swath of East Africa that extends from Kenya down to Mozambique—was infamous for ports where Arab traders would take captives from the interior and ship them around the Arabian Sea and farther to countries on the Indian Ocean. On the western side of the continent, the slave trade began in earnest with the Portuguese taking captives from the kingdoms of Angola and Kongo, and ravaging a wide swath of the country along the west coast. But the centre, the Great Lakes region, was relatively untouched because it lay beyond the easy reach of slave traders operating on the coast, and because it had strong political systems that kept intruders out.

When I began to learn about what Central Africa was like five or six hundred years ago, I thought of the hodgepodge of political groupings—called duchies or counties or principalities and sometimes kingdoms—that governed Europe for several centuries following the fall of Rome. In Central Africa these units were the equivalent of kingdoms, and were led by men who wielded considerable power over large areas. Their existence had a profound effect on the territories that would become Rwanda and Burundi, although the European powers who wanted to divvy up Africa in the 1880s knew very little about the politics of the continent or even its geography.

Take a look at a map of Africa from the middle of the nineteenth century: the lack of detail in the middle is striking. Europeans knew that the Nile had its headwaters somewhere in the mountains south of Egypt, but no one had plotted that legendary river's course. Nor did they know where the Congo, the great river of the western coast of Africa, began. Nevertheless, Belgium, Germany, France, Italy, Portugal, Russia, Spain, and the United Kingdom all claimed and received "international"—meaning European—approval for controlling the continent. Some of these countries had long-standing claims to parts of Africa, like Spain and Italy in the Maghreb and Portugal on the east and west coasts. But elsewhere the process was rather like a skulk of foxes dividing up rights to a newly discovered henhouse. As the *New York Times* editorialized in 1964: "Then came the European mapmakers, drawing lines and spaces on sheets of paper which they duly colored and named. All of Africa today is a patchwork of tribes living on territories that often have no relation to political frontiers."[13]

The kingdoms of Rwanda and Burundi were originally allotted to the Germans, who ruled indirectly, co-opting local royalty to run things. But during World War I the Belgians—whose King Leopold had set up his own private empire next door in the Congo—acquired the kingdoms, and subsequently governed them under a League of Nations mandate. Ironically, grouping the kingdoms with Congo made a certain amount of sense, since the Bantu people who lived there had much in common and the ties of power between local royalties frequently transcended the boundaries drawn up in Europe. But after the war the Belgians decided to govern the Congo and what they called Ruanda-Urundi as two separate entities.

A word should be said about the kingdoms' ages, which are, notes historian René Lemarchand, "a striking anomaly on a continent of artificially fashioned state systems."[14] Burundi, he stresses, existed as a national entity long before many European states, and its boundaries have remained "virtually unchanged" since the mid-nineteenth century: the borders between it and Rwanda are not mere lines drawn on a map, but the course of two rivers, the Akanyaru and the Kagera.

In Rwanda an independent monarchy appears to have existed for even longer than Burundi's. One important difference must be recognized though: in Burundi the kingly caste was considered above both Tutsis and Hutus, claimed to have a mixed ancestry, and governed with the advice of a council of sages drawn from all over the country; the Rwandan royal family, by contrast, was completely Tutsi, and Tutsis had exercised a monopoly of power for generations.[15]

As noted, without exception European colonizers were certain that Tutsis were superior to Hutus, even though it was frequently hard to tell to which group an individual belonged. To avoid making the embarrassing mistake of mixing the two up, the Belgians—who administered the territory of Ruanda-Urundi from Bujumbura, a new city they built on the shores of Lake Tanganyika—instituted a system of identity cards. It was supposedly based on "objective" characteristics such as height or the shape of one's nose and lips, but there was such physical overlap in the population that, in implicit admission of how subjective the categories were, the ownership of cattle was added as a criterion in the late 1930s: anyone with more than ten cattle was henceforth classified Tutsi.

As late as 1931 observers noted very little difference in the material well-being of the two groups. Catholic missionary Bernard Zure wrote then that the differences in attitude and behaviour between Hutus and Tutsi "have become so minimal that one can speak of a common culture."[16] By 1950, the average income of Tutsis and Hutus was about the same and a small elite of Hutu teachers and businessmen had developed.[17] Nevertheless, Belgians were so wedded to the idea that the Tutsis were intrinsically higher class that they grossly underestimated their numbers. In 1954 the Tutsi were estimated to account for only 7 percent of the entire population, but when a real census was done two years later, the figures were much higher: between 13 and 18 percent, depending on where in the region the count was made. It's just that poor Tutsis had been invisible to Europeans.[18]

Until the 1950s, the Belgians, with the support of the Roman Catholic clergy, systematically put Tutsis in positions of influence. When they revamped the administrative units by merging some and thus cutting down on the numbers of sub-chiefs, it was Hutus who lost their posts, not the Tutsis. Tutsi youth also got preferential admission to schools because it was believed the Hutus would not profit from a good education.

Despite this bias on the part of the colonial administrators, overt conflict between Hutus and Tutsis was uncommon until the 1950s, when the winds of independence began to sweep through Africa. Following World War II, Belgium administered Ruanda-Urundi as a United Nations Trust Territory and was supposed to prepare it for independence. The small country dragged its feet, however, in part because it was reluctant to give up its status as a colonial power.

The first violence of note between Hutus and Tutsis occurred three years before the Belgian trust became two separate countries. To protest growing

Hutu demands that they have power in keeping with their numbers, in 1959 supporters of a pro-Tutsi party attacked a popular Hutu chief in the west of what is now Rwanda. The man survived, eventually becoming the first president of the independent Rwanda, but rumours that he was dead ran wild at the time. In reprisal Hutu militants killed Tutsi civilians, and tens of thousands of Tutsis fled to Burundi, Uganda, and Tanzania.[19]

The repercussions of that conflict reverberate today. Among the Tutsi exiles was the family of two-year-old Paul Kagame, who would be educated in English in Uganda and return at the head of the Rwandan Patriotic Front after the 1994 genocide. Note also that the Hutu *génocidaires* of 1994 said they did their "work" (the code word used in the call to attack) in memory of "1959."

Initially, the Belgian plan for independence was to have both countries be constitutional monarchies, but in short order that was abandoned in Rwanda. The king there was far more tightly identified with the Tutsis than his counterpart in Burundi, and Rwandan Hutus clamoured for a republic from the beginning. It was at this point that a change in attitude occurred among the Belgian authorities, who began to think that giving power to the Hutu majority was a good idea. Some observers say this was either a conscious or unconscious reflection of the conflict playing out at home between French-speaking and Flemish-speaking Belgians. Be that as it may, when the plans for independence were made, it was clear that Rwanda would be a republic, with the majority ruling.

In Burundi, however, the oldest son of the king was extremely popular. A Tutsi, Prince Louis Ragaswore also had a burly, Hutu-like physique plus great charm and intelligence. His party won a massive mandate in the first parliamentary elections in September 1962 despite the fact that in Burundi, as in Rwanda, a good 80 percent of the population was Hutu. For a time—a very short time—it looked as if there might be a government that engaged all segments of the country, and could get them to work together.[20]

However, two weeks later the prince was shot dead as he lunched on the terrace of a lakeside club. The killer was a Greek national, but those behind the murder were aligned with Belgian forces who didn't want to let power slip into Tutsi hands. Departing Belgian officials called Ragaswore "stupid, conceited, spendthrift and party-going." They charged that his political party was anti-Belgian and "dangerously pro-Communist."[21] Given the Cold War politics of the time, charges like that were a red flag to other Western powers. Just how much the United States meddled is unclear, although events were followed closely by correspondents of the *New York Times*, which suggests official interest.[22] So does the interest that the United States and Cuba showed in the Rwandan refugees living in Uganda during the 1960s and '70s.

One of them was Paul Kagame, who fought with Ugandan rebel forces during the conflict that led to the ouster of Ugandan dictator Idi Amin in 1979.[23] Following that victory, the new Ugandan leader, Yoweri Musevani, sent Kagame to Cuba with sixty-seven intelligence officers under his command for training. Then—also at Musevani's bidding—he spent a few years at Fort Leavenworth, Kansas, where he studied at the US Army Command and Staff College. When he came back to Africa he threw himself into planning an invasion of Rwanda by Tutsi exiles. By then, though, the outside world was much less concerned about Central Africa, because the Cold War between the United States and its allies and the Communist Bloc ended at the beginning of the 1990s. What was happening there was no longer on the Western world's radar, and when conflict erupted in Rwanda the news was received with great surprise even though the signs had been there for those who wanted to see.

That lay thirty years in the future when Burundi's Ragaswore was assassinated, however. Yet his memory, and regret over the loss of what might have been under his leadership, lives on in Burundi.

Not a good start: Prince Louis Ragaswore was assassinated on the terrace of the Club Tanganyika in 1962, shortly after his party won a massive victory in Burundi's first parliamentary elections. *Photo provided by author.*

That became startlingly clear to me when I got a guided tour of Bujumbura from a young Tutsi woman who worked for an NGO that at the time was doing the groundwork to help provide land title for farmers.[24] The daughter of a well-placed businessman whom I'd met on the flight from Nairobi to Bujumbura, she took me first to the monument to Ragaswore's memory, high on the hills that I had looked out onto earlier that morning. Not only was the view to the west across Lake Tanganyika gorgeous, it was clear that Ragaswore's tomb was a shrine of great importance. I had no idea just how powerful his legacy was, however, until a few days later when the young woman's father took me to lunch at a restaurant on the shores of the lake.

The day was hot, with the sun trying to shine through the humidity; a thunderstorm was promised for later in the day. We talked about many things as we ate: the city, his work, his family, my family, and a bit about the prospects for peace. When the owner came out to say hello, my companion introduced me, and I sent my compliments to the chef: the meal was the best that I'd had in Burundi. In short, it was as pleasant an interlude as could be expected on a trip where you're trying to find out some truths about a part of the world that has been shaken with paroxysms of violence.

Then we got up to leave, and as we crossed the terrace toward the white-pillared entrance and the lawn leading down to the lake, my friend pointed off to his right, to the spot where Prince Louis was gunned down. Things might have been different had the prince lived, he said, and there was great regret in his voice.

— — —

After Ragaswore's death there were two major bloodbaths in Burundi, which some have qualified as genocides, one against the Hutus in 1972 and one against the Tutsis in 1993. The Hutu majority in Burundi was the victim of smaller massacres on several occasions—in 1965, 1969, 1972, 1988, and 1991.[25] Nothing, however, compares with the violence that occurred in Rwanda in 1994. When an airplane carrying the presidents of Burundi and Rwanda was shot down that April, it was not immediately obvious to the outside world what was happening. The UN forces that had supposedly been monitoring the tense relations between Hutus and Tutsis sent out warnings that were not heeded: the UN commander on the ground, Canadian Roméo Dallaire, was told to stay neutral when trouble broke out. As things went from bad to worse, the world watched in horror but did nothing. The violence continued until nearly all Tutsis and moderate Hutus in Rwanda were killed or in exile. It ended only when Kagame's Rwandan Patriotic Front marched in from Uganda and took over.[26]

It is at this point that the paths taken by the two countries began to diverge, when the unidentical twins headed in directions markedly different.

Following the 1993 Tutsi attack on Hutus in Burundi, an international commission set about investigating what had happened.[27] Its findings are sobering, but perhaps the most disturbing is the conclusion about what should happen next:

> It must be borne in mind that, among the present adult population of Burundi, tens, if not hundreds, of thousands of individuals from both ethnic groups have at one time or another committed homicide. To prosecute every one of them is clearly beyond any system of justice. If those who bear the main responsibility for these crimes are ever to be brought to justice, judges or prosecutors must be empowered to offer immunity or reduced sentences, in exchange for cooperation, to those who were merely ordered or led.
>
> The establishment of an impartial and effective system of justice would require considerable international assistance in the form of training and financial support. A transitional period could be envisaged during which, to establish public credibility, foreign observers, recruited from the judiciary of other francophone African nations, would sit in the bi-ethnic courts and, if necessary, mediate between the judges.

But nothing came of that. There were no trials, no official attempts at peace and reconciliation, no help from foreign judicial experts. It was only in 2015 that the unofficial Russell Tribunal—initially organized by philosophers Bertrand Russell and Jean-Paul Sartre to investigate war crimes committed by the United States in Vietnam—set about trying to determine if crimes against humanity were indeed committed twelve years before in Burundi.[28] The tribunal's findings—whenever they are handed down, and it is by no means clear when that will be—will carry no force of law, just of public opinion.

But in Rwanda, perhaps because the scale of the carnage was almost unbelievable, an approach somewhat similar to that proposed for Burundi was actually undertaken. Two levels of justice were set up. On the local level, *gacaca* (meaning "justice amongst the grass") courts were established to take care of the thousands and thousands of cases of people accused of genocide and war crimes. Based on a traditional Rwandan institution, they were designed "to restore order and harmony within communities by acknowledging wrongs and having justice restored to those who were victims."[29]

It was a tall order: in the end, 12,000 community-based courts with locally elected judges tried more than 1.2 million cases. They frequently gave shorter sentences if the accused confessed and was repentant. The national court system, which, without the local courts, would have been completely overwhelmed by the number of accused, handled the cases of about 10,000 persons accused of planning the genocide or committing serious atrocities. The most serious offences were heard by the International Criminal Tribunal between 1997 and 2012, which sentenced 61 people to terms of up to life imprisonment: the death penalty had been abolished in 2007.[30]

Just how important the world thought these courts were must be measured against what happened to the accused in other genocides and ethnic cleansing campaigns. While the UN tribunal was at work in Arusha, war criminals from the conflicts in the former Yugoslavia were being tried at The Hague in the Vredespaleis (Peace Palace), an imposing Neo-Renaissance building constructed in 1913 for the International Court of Arbitration, which was created

The International Criminal Tribunal for Rwanda heard the most serious offences in the 1994 genocide in Rwanda between 1997 and 2012, and sentenced sixty-one people to terms of up to life imprisonment. *Source: ICTR, public domain.*

in 1899 to "end war." Set in formal gardens and surrounded by an ornate fence, the building is impressive and not uninviting—until you try to gain access. Clearly, all visitors must be vetted. On a stopover on my way to Africa I was turned back by guards: there was no way a person off the street could get in just because she'd like to take a look.

Things were much different in Arusha, where the court was housed in a tall, modernist building like thousands found around the world. All it took was a query at the desk to be admitted. Yes, there'd be no problem with a Canadian writer observing the trial in process that morning if she left her passport as surety. The session would begin in half an hour, and in the meantime she could wait in the library, read the *International Herald Tribune* (a couple of days late), check her email, read the pamphlets prepared for possible witnesses. Would you like a cup of coffee too? asked one of the staff members, a young Tanzanian.

When it came time for the session to begin, he shepherded me to the floor where the entrance to the press and visitors' gallery was. We had to take a particular elevator. The witness to be heard took another, and curtains protected her from view in the hallway outside the courtroom and in the witness box.

On trial was a Seventh-day Adventist pastor who was also head of a hospital where the woman testifying had worked as a nurse. She spoke Kinyarwanda, which was translated into French and then into English so all the lawyers and judges could understand. During the troubles she said she had hid in the bush for a few days and then went to the hospital with her children, believing that they would be safe there. But they and the others hiding there—mostly Tutsi, but a few Hutus married to Tutsis—weren't.[31]

Just what happened to them I didn't hear that day, because she became ill; whether from the stress of confronting the man who wanted her dead (he was later found guilty), or due to a physical cause, wasn't clear. I left the courtroom shaken, almost overwhelmed by what I'd heard, and by the thought that eight years later the very sight of a man might incapacitate one of his would-be victims. Clearly the shadow of the genocide would lie long on this land.

— — —

This judicial process seems to have worked well enough so that Kagame's regime has been able to control conflict: there are practically no reported incidents of ethnic violence even though (or because) mentioning the terms "Tutsi" or "Hutu" is now illegal.[32] It looks also as if Kagame will be in the saddle for some time to come. After a constitutional amendment allowing him to run for a third term was approved, he did, winning handily with 93 percent of the vote in August 2017. His run was undertaken reluctantly, he says, but another seven-year term (and two possible five-year terms after that) will allow him to consolidate the gains the country has achieved under his rule, he and his supporters contend. Critics counter that during the twenty-three years that Kagame's party has been in power opposition groups and independent voices were systematically silenced. The election was held in a climate of fear, they say, so its results aren't valid.[33]

Kagame's aim, he says, is to create the African equivalent of Singapore—that is, an economically prosperous state with good education, health, and financial systems. Certainly the contrast now between the good roads and booming construction in Rwanda and the situation in Burundi is remarked upon by all who visit the two countries.

"Travellers in Africa are always taken aback at how *tidy* Kigali is," Canadian writer Will Ferguson noted in his book *Road Trip Rwanda: A Journey into the New Heart of Africa*:

> Glass towers are spinning themselves into existence on the dizzying pirouette of construction cranes, but even with the city's population topping a million and growing daily, there are no sprawling slums, no shantytown ghettos stretching in the distance, no garbage pickers living on smouldering hills of trash. This is urban Africa reimagined.[34]

Whereas in Bujumbura, far too much has remained the same or gotten worse, suggests Gaël Faye in his novel *Petit pays*. The son of a French father and a *Rwandaise* mother exiled in Burundi during the troubles of the 1980s, Faye describes the Burundian capital of his childhood as a paradise lost. While he insists the novel is not really autobiographical, he writes movingly of what happened in the 1990s and of his return after years in France, where both he and his hero Gabriel found refuge during the conflict:

> Twenty years later, the *cul de sac* where we lived had changed. The big trees of the neighbourhood had been chopped down. The sun overwhelmed the days. Concrete block walls topped with crushed bottles and barbed wire had replaced the colourful bougainvillea hedges. The impasse was no more than a sad, dusty corridor, in which its anonymous inhabitants were confined.[35]

Nevertheless, the conversations of the drinkers in the neighbourhood bar hadn't changed:

> They talk of upcoming elections, peace accords, the fear of a new civil war, their unhappy love life, the increase in the price of fuel and sugar. They pour forth the same hopes as before, the same rants. The only new thing is that I hear the ring of a cell phone from time to time, to remind me that the times have indeed changed.

But note that while Faye wrote his novel in French while living in France, over the last decade French has become much less important in Rwanda than it once was. Kinyarwanda is still an official language, but English has replaced French as the second language. Young people are supposed to be educated in English beyond the first grades, and, while there is an acute shortage of teachers competent in English, it's clear that Kagame—who was educated in English himself in Uganda, remember—wants closer ties with the English-speaking powers. He's wrangled membership for Rwanda in the British Commonwealth of Nations, although he's also jostled for prominence in the Organisation internationale de la Francophonie. With vigorous support from French president Emmanuel Macron, Kagame put forward Rwanda's foreign minister for secretary-general of the organization, pushing aside the incumbent, former Canadian governor-general Michaëlle Jean, whom we met in the chapter about Haiti and the Dominican Republic.[36]

Burundi was represented at the Francophonie summit but appeared not to play an important role. Back at home, a power-sharing agreement, achieved after ten years of civil war, appeared to be unravelling. Pierre Nkurunziza, a burly Hutu who plays soccer seriously and owns his own team, was elected by the parliament to be president in 2005. His government contained both Hutu and Tutsi members, while under the peace agreement the military had an ethnic split of 60 percent Hutu and 40 percent Tutsi. But Nkurunziza's government has not been noted for good governance. On several international comparisons, Burundi is rated as one of the world's most corrupt nations, while its per capita income has stagnated near the bottom.[37]

When the time drew near for Nkurunziza to step down as president after the two terms Burundi's constitution allowed, he argued that the restriction didn't apply to him since he had been elected only once by popular vote. Nkurunziza has said in the past that he believes God has destined him to govern Burundi, so perhaps it's not surprising that he ran again, winning handily. And he might run again in 2020, because a constitutional referendum passed in early 2018 gave him a green light to do so. He's said since that he will step aside and allow for a transition in power, but many are skeptical about his intentions.[38] Certainly the country has been in turmoil: in July 2017 the International Federation for Human Rights warned that the country was on the edge of a precipice, with 1,200 dead, hundreds of forced deportations, thousands tortured, and 400,000 refugees.[39] There appears to be no evidence that things have gotten better since.

As for moving closer to the Anglophone world, Nkurunziza has shown no interest. Kirundi and French are the official languages, with education, such as it is, being conducted in Kirundi at beginning levels and French for more advanced study.

Yet despite the differing paths Burundi and Rwanda have begun to take, some important similarities remain. Most importantly, demographic pressures didn't let up after the great spasms of violence in the 1990s. The population in both countries has continued to increase with similar rates of growth. This means that despite the number of people killed or exiled, population density in both countries is actually higher than it was when the violence exploded.

What will happen next is anybody's guess. Perhaps it is significant that ten years after my trip to Burundi, the youngest daughter in the family that had been so kind to me in Bujumbura arrived in Montreal to study. She didn't choose one of Quebec's fine French-language universities, but rather McGill. She now has her feet in two worlds, fluent in English and French, but is not terribly interested in going back home. Those who can get out, do.

SCOTLAND
AND IRELAND

When you can, you get out, indeed.

That's what my grandfathers did. They ran away in their teens and neither looked back. Their reasons were complicated and I only partially understand why they did what they did, but I do know that, just as my young friend from Burundi was escaping the consequences of colonialism gone horribly wrong, my grandfathers' flights were episodes in a long colonial drama.

Yes, colonialism. It has played a role in all the stories we've considered so far. What to do about its legacy was a key question of the twentieth century, and its shadow is long in the twenty-first. In the case of my grandfathers, the colonial power was England, and the story goes back generations before they were born. In the case of my husband and myself, the long arm of British colonialism was a factor, but so was the modern colonialism of the United States.

But first a digression, which may not seem relevant but is. The second winter after we arrived in Montreal, we still had the stalwart vw Beetle that we drove across the continent in that summer of love and contention. We didn't use it very much, though, because we were living in the centre of town. All to the good in the winter, because vw Beetles were notoriously cold—later ones had gasoline heaters, but ours had none. Yet occasionally I had to drive in the snow. On the day I'm remembering I manoeuvred the car out of the apartment garage and into the laneway through the remnants of a half-metre of snow that had fallen two days before. Beetles also had a reputation for handling well in snow because the engine was in back, over the traction wheels. Perhaps that had made me a little overconfident, too sure that I could get through the snowbank blocking the entrance to the street. *Gun it*, I said to myself, and the gutsy little machine roared forward, only to stop two feet into the bank.

For a moment I sat there and reviewed my options. I'd never shovelled much snow, but how difficult could it be? Back on the West Coast, I wasn't considered a big woman, but here I'd seen several women much smaller than me clearing walkways. And I was young: mid-twenties, prime of life. So I got out, took the shovel from the trunk—which was in the front in those old-fashioned, engine-in-the-rear vws—and attacked the snow.

There are two things you should know at this point. One is that, while I had been studying French ever since we arrived in Montreal fifteen months before, my grasp of the language was still spotty. I could read it all right, even without a dictionary, but understanding spoken French was something else. The other thing is that I had a great mop of red hair. Back in San Diego, this was very unusual. In my Southern California high school class of five hundred there was only one other redhead, and she was probably a shirttail relation by way of the Scottish Highlands since her last name was Macdonald and my mother's maiden name was McDonald: different spelling but same clan, I'd been led to believe. Therefore I assumed that anyone with red hair spoke English.

Okay, so there I was with my hair whipping around in the wind, and the snow flying in all directions. I worked for about five minutes, but when I stopped to take stock I could see that it would be a long time before I could get my path clear. *Zut!* I said aloud: it was a new term that I'd adopted from the latest dialogue I was supposed to memorize for my French class. *Zut!* How the hell am I ever gonna get this car out?

Then my saviours arrived: a City of Montreal snow-clearing crew. Five men with a snowplow attached to a truck, and a sidewalk tractor. They stopped and watched for a couple of minutes, and then the boss shouted something at me through the truck window.

"Got a problem?" is probably what he said, but it was in French and in my flustered state I hadn't a clue. I straightened up and looked at the grinning guys. For a second I wondered if I should ignore them; that was the technique I found best to avoid being hassled by men. But these guys radiated such amused good will! This wasn't harassment, it was help.

That's when I spied the youngest fellow, standing up in the back of the truck. His hair was the same riotous red as mine, so I figured he would understand me. We had to be related somehow, he should know English.

"No understand," he said, but he jumped down from the truck and started to shovel anyway. Two of his co-workers helped. In a very few minutes they had cleared my passage. Then, with signs, they told me to get back in the car, and positioned themselves so they could push if I got stuck again.

I didn't but I stopped once the car was in the street. "Thank you," I said through my open window. "Merci beaucoup."

My red-haired friend waved away my thanks.

"What's your name?" I called out. "Comment vous vous appelez?" I dredged up from another dialogue that was included in my French workbook.

"Marc-Antoine O'Neill," he answered. Irish, not Scotch (as my grandfathers called themselves.) A descendant of one of the many Irish Catholics who had fetched up in Quebec in the nineteenth century. Many married into French-Canadian families, others were immigrant children who'd been orphaned by accident or disease and subsequently adopted by a Francophone Catholic family.[1]

And that's when I began to get a fix on how complicated the great Celtic web was, how interrelated are the destinies of Scotland and Ireland, the next pair of unidentical twins, of frenemy nations, we'll consider here.

There was a period, back in the dawn of time or almost, when all of the British Isles were part of the main, when there was no distinction between Ireland and Scotland.

For thousands of years, the region—along with much of the rest of Europe and North America—had been covered by ice, kilometres thick in places. Then, as the last great Ice Age ended, humans began moving into the newly found land, beginning about 15,000 BCE. Because so much of the earth's water was locked up in ice, the oceans were smaller, and sea levels much lower. If you'd been around, you could have walked from France to Galway.

As the ice retreated, the region became a haven for hunters and gatherers. Evidence for this has been dredged up for a hundred years by scientists aided by fishermen in an area of the North Sea between Britain and the European coast called the Dogger Banks. (*Doggers* is a Dutch term for fishing boats.[2]) Humans lived rather successfully there for a few thousand years until Doggerland, as it is now called, was submerged by rising sea levels as the Ice Age ended. The final blow came about 6200 BCE when the earth's crust rebounded, freed from the weight of the ice cap. An earthquake and landslide near Norway caused a catastrophic tsunami with waves 4.5 metres high, flooding the region and definitively cutting the British Isles off from Europe in geographic terms.[3]

But human life continued on the islands, life that was far from primitive. While little has been found that dates to the epoch of the tsunami, impressive stone temples and monuments were built about 3200 BCE in both Ireland and on the Orkney Islands, off the northwest coast of Scotland. Newgrange, a mammoth temple complex in northeastern Ireland—today a UNESCO World Heritage Site—was built at least five hundred years before Stonehenge in England or the Giza pyramid in Egypt. It has fascinated since its accidental discovery in the seventeenth century, when the owner excavated what appeared to be a hillock in order to get stone blocks for building barns, or granges.[4]

What was found was apparently the seat of religious practice and burials. Deep inside there is a chamber that the sun lights up for slightly more than a quarter-hour at the winter solstice just after sunrise, suggesting both sophisticated astronomical knowledge and rich religious rituals. A circle of massive stone uprights surround the main building, bearing witness to the technical skill of the people who cut and transported them without the aid of any metal tools.

Other monumental constructions on the Orkneys at Ness of Brodgar were uncovered much more recently, but they have also become a UNESCO World Heritage Site. They've been compared to the Acropolis in Greece because of the way they were built to dominate the landscape, but they are twenty-five hundred years older than the classic Greek structures. Not far away is a tomb mound called Maes Howe, dated to 2500 BCE, the inner chamber of which is lit briefly during the winter solstice, just like at Newgrange.

The remains of a Stone Age village called Skara Brae are also nearby. Dating from 3100 BCE and apparently occupied for more than six hundred years, it was uncovered in 1850 when a wicked winter storm scoured sand away from the ruins. The complex of houses and farm buildings give an unequalled glimpse of what life was like in the flourishing communities of the northern British Isles at a time when agriculture was just arriving in that corner of the world.

Who these people were and what ended their period of glory are unresolved mysteries. One of the surprising things about the Ness of Brodgar monuments is that they did not slowly fall to ruin from neglect. Instead, the temple was "decommissioned and partially destroyed," as Nick Card, excavation director with the Archeology Institute at the University of the Highlands and Islands, told *National Geographic*.[5] The remains of four hundred cattle, all slaughtered at the same time, were placed in a circle around the main structure, and then the buildings were demolished and buried under tons of rubble. "It seems that they were attempting to erase the site and its importance from memory, perhaps to mark the introduction of new belief systems," Card says.[6]

The people who built Newgrange and the Ness must be classified as Stone Age despite their great skill, since the secrets of working metal had not yet made it to the islands. The end of their culture could be due to changes in climate that made northern sea lanes more dangerous, thus displacing trade routes to the south. Crops formerly successful in the area might also have been affected by different climatic patterns.

But that is definitely not the whole story. Archeological and genetic evidence point toward massive population and cultural shifts that brought new people as well as techniques for working bronze and then iron to the islands. Like the Bantu who expanded from the interior of western Africa into the central and eastern portions of that continent with a tool kit containing iron-working skills and new crops about the same time, the newcomers in Europe carried with them language and skills that apparently overwhelmed the indigenous populations.

— — —

Trying to figure out what happened back then is rather like trying to knit with telephone poles. But advances in genetics help, because looking at the genetic makeup of individuals now and comparing that with DNA from the remains of humans hundreds, even thousands, of years old can tell a lot about population movements. Sequences of DNA can now be analyzed, uncovering variations in genes, called alleles. Scientists have done just that with the remains of four people who died millennia ago in what is now Ireland. One, a woman farmer found near Belfast, died about the time that Newgrange was in its prime: indeed, her remains were buried in a passage-like grave, reminiscent of Newgrange. Her DNA suggests she had black hair and brown eyes and that her ancestors came from near the Mediterranean, making it likely that they brought with them knowledge of how to grow things.[7]

The other three were men who lived and died fifteen hundred years or so later than she did. Their remains were recovered from cist burials—small stone enclosures that contain the bones of one or a few individuals, not the many that are found in passage graves. At least one of them may have had blue eyes, a recessive trait that requires two pairs of the gene that programs for the characteristic, while the other two had one gene for the trait.[8] The scientists involved in this study say these men's ancestors probably came from as far away as the steppes of southern Russia. An even more extensive study of the genetic makeup of the ancient remains of four hundred people from across Europe suggests that this population movement was the end of a reverse "Neolithic Brexit," as one of the scientists involved told the BBC in early 2018.[9] For a good fifteen hundred years there appears to have been little contact between the populations of the British Isles and the European mainland, but within a few hundred years newcomers replaced the earlier agricultural population.

This migration brought with it knowledge of how to work metal as well as, apparently, genes for digesting milk after infancy and for a blood disorder called haemochroataosis, which is common today in Ireland. The former characteristic, you'll remember from the discussion of lactose tolerance (also called lactase persistence) in Burundi and Rwanda, is practically universal in Ireland today.[10]

That the young men who lived in the later period have light hair and skin says a lot about what was happening in Europe back then. When humans set out from Africa to people the rest of the world, their skin was almost surely dark. As we saw in the chapter about the Indian states of Kerala and Tamil Nadu, dark skin provides some protection from ultraviolet rays from the sun, so it's not surprising that dark skin gave people living near the equator an evolutionary advantage.

However, when humans began to move into territory where day length varied, and colder temperatures required covering up to keep warm, light skin was advantageous because it produces more Vitamin D—sometimes called the "sunshine vitamin"—than does dark skin. Not having enough of the vitamin can lead to crippling bone malformations that, among other things, can make giving birth extremely difficult if not impossible. The power of this selective pressure is such that light skin seems to have been selected for at least twice in the history of humanity, once among Europeans and again among East Asians, though each were the result of a different gene form.[11]

The picture of where these changes took place is being sketched by scientists. At the moment, it appears that much of the evolution took place in the vast steppes abutting Eastern Europe, where the horse was domesticated.

At some point this culture expanded westward as well as eastward. It was so successful that it appears to have fundamentally changed most of the communities it encountered, either by conquering them or by simply showing them a more productive way of life. The trio of dead Irishmen must have been among them, as were the people who arrived in what is now Scotland.

Note that none of the trio appear to have had red hair. Now, if you're looking for things that Ireland and Scotland share, a lot of redheads is one of them. The trait is far more common there than elsewhere in the world, except for an area in Russia along the Volga River. It's estimated that perhaps 1 percent of all Europeans are redheads, but among the Irish the figure may be 10 percent, while in Scotland the percentage could be about the same. The exact numbers of redheads in each country aren't known, although DNA projects are trying to discover how many people carry the recessive gene that produces red hair.[12] BritainsDNA, a commercial DNA testing firm, recently tested a sample of 2,343 people whose grandparents all came from England, Scotland, Wales, or Ireland. Their results showed the southeast of Scotland, particularly around Edinburgh, as the hot spot, with 40 percent of the sample there carrying one of three common red hair gene variants.[13] Not all of these people were redheads—you have to inherit the variant from both your parents to be carrot-topped—although a great many are.

Yet the British Isles are not where red hair originated. Several references to red hair are found in ancient texts with origins far from there. Herodotus, writing in the fifth century BCE, speaks of the red-haired Thracians from the Greek peninsula and present-day Turkey, and the red-haired Budini from east of the Don and north of the Black Sea.[14] But the earliest physical evidence of red hair goes back much earlier to what is now western China. There, a number of mummified corpses dating back to 2000 BCE show distinct "European" facial features and red hair.[15] What's more, some of the mummies are clad in textiles that look very much like the plaids woven by early Celtic artisans in what is now Austria.[16]

The Celts? Well, of course, both Ireland and Scotland are identified with the Celts, and plaids loom large in Scots and Irish traditional dress. Irish and Scots Gaelic are both Celtic languages too, and there is ample evidence that Celtic mythology and culture held sway in both countries for hundreds of years, or longer. Indeed, it could be said that Ireland and Scotland are the outliers in a great colonizing movement that left practically no written records of emperors or high kings, but which nevertheless ought to be compared to the other great powers of the ancient world.

Like the Phoenicians, who were more a loose confederation of trading cities than an empire headed by Great Men, the Celts appear to have colonized as much as conquered. Also like the Phoenicians (but for different reasons), few documents or monuments lauding their accomplishments have survived. Julius Caesar, who campaigned in Gaul and invaded England in 58–59 BCE, noted that the Celts or Gauls (more on these terms in a moment) used the Greek alphabet to transcribe information about transactions and population data, but that Celtic poetry—which presumably would contain the stories of heroes and conquerors—was committed to memory rather than written down.[17] Aside from a few stone markers, these documents have been lost, although archeological finds of burial grounds at Hallstatt in Austria and at La Tène in Switzerland fill in the record. Salt mining began at Hallstatt somewhere between 800 and 500 BCE, bringing great wealth to the community, and treasures found in graves there bear witness to a flourishing Celtic culture.[18]

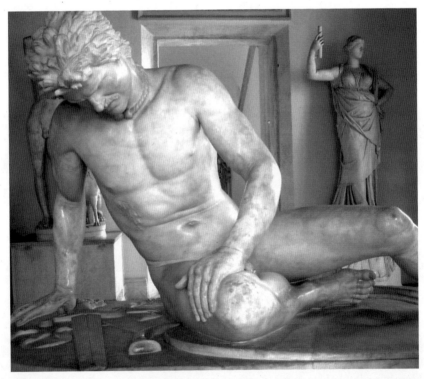

The Dying Gaul: a mustachioed warrior with a Celtic ornament—a torque—around his neck, is a portrait of a warrior that the Greeks and Romans considered an admirable adversary. *Source: Creative Commons Attribution-Share Alike 3.0 Unported licence.*

The first references to Celts in Greek writings date from about the same period. The Greeks called them *Keltoi* (sometimes construed as meaning "The Tall Ones"), but many of them called themselves Galls or Gauls. The extent of their conquests show up in place names like Galicia in the northwest corner of the Iberian Peninsula and Galatia in what is now Turkey. They were formidable fighters, sacking Rome in 379 BCE and battling the Greeks in 291 BCE at Thermopylae when they tried to move south and east. The Greeks considered them admirable opponents. A statue of the Dying Gaul—first thought to depict a wounded gladiator—testifies to that. The existing marble is a Roman copy of a lost Hellenistic statue that shows a mustachioed warrior with a signature Celtic ornament called a torque or torc around his neck. Made of twisted metal wire or a thin metal bar, torcs were worn by Celts across a wide expanse of Europe for centuries if not millennia. The man who wears it here is depicted as a valiant adversary, showing "courage in defeat, composure in the face of death, and dignity"—all things considered virtues by Greeks and Romans.[19]

But Celtic, or Gallic, religion was dreadful. Julius Caesar wrote that human sacrifice and cannibalism were commonplace.[20] Strabo the Geographer repeated the same claims.[21] In modern times scholars have been loath to give credence to these tales—one calls them war propaganda—but the recent discovery of a young man, garroted and with his throat slit with what appears to be great ceremony, gives credence to the belief.[22] So does a study published in August 2017 showing ritual carving and human teeth marks on human bones dated to fifteen thousand years ago (plus or minus a couple of millennia) in Britain.[23] But before one becomes too repulsed by the idea, consider that a number of other civilizations and societies have practised ritual human sacrifice. An echo of the custom may be seen in the biblical story of Abraham, who was ready to sacrifice his son Isaac until a lamb miraculously appeared as a substitute. And then there is the Christian story of God sacrificing his only son for the salvation of mankind...

Indeed, it's this powerful narrative at the heart of Christianity that convinced the Irish to convert, asserts Thomas Cahill in his *How the Irish Saved Civilization*. Because the supreme sacrifice had already been made, there was no need for further sacrifices of humans or animals. Saint Patrick, a Romanized Briton kidnapped and sold into slavery in Ireland, preached this gospel beginning about 432 CE to people whose spiritual beliefs were dramatically different.[24] The result of this evangelizing was the establishment of the church in Ireland in the years that saw Britain descend into chaos after the Roman Empire collapsed.

Christianity would become one of the two things that united Ireland and Scotland, building on top of the heritage of Celtic culture. Language was the other. Gaelic was their common tongue for centuries, and even when the Royal Court of Scotland adopted a dialect of Middle English called Scots in the twelfth century, the link of language lived on.[25] Two hundred years later the Scottish hero Robert the Bruce was writing to his Irish friends in an attempt to convince them to make common cause against the English, who were attacking Scottish borders:

> Whereas we and you and our people and your people, free since ancient times, share the same national ancestry and are urged to come together more eagerly and joyfully in friendship by a common language and by common custom, we have sent you our beloved kinsmen...to negotiate with you in our name about permanently strengthening and maintaining inviolate the special friendship between us and you, so that with God's will our nation...may be able to recover the ancient liberty.[26]

The alliance came to nought but religion continued to bind the two peoples together.

That may sound surprising, given the antagonism that later developed between the Protestants of Scotland and the Roman Catholics of Ireland, and which to some extent continues today. But back in the fifth century Saint Patrick's Ireland led the way toward Christianizing the pagan Scots, who had repeatedly pushed back Roman advances, and so had not become Christian when Britain, along with the rest of the Roman Empire, officially adopted that faith in 300 CE.

Ireland was never conquered by the Romans either. Both it and Scotland were competent metal-working societies that faced a civilization that had more sophisticated techniques for organizing and controlling people, goods, and production. By no means, however, was the clash of cultures one-sided. Roman law is justly famous, and influences much of the Western world to this day. But the Irish legal tradition was far from primitive: even before they had adopted a writing system, the Irish had their own system of laws administered by judges, or *Behans*. Later codified by Saint Patrick in the five-volume *Senchus Mor*, they allowed nearly everyone in Gaelic Ireland and Scotland to participate in the selection of tribal leaders or kings. Women also wielded considerable clout, both literally and figuratively. They had a right to be educated and to own property, while elected queens led armies into battle. Even after Christianity arrived, and Christian practices in 697 CE mandated women's exemption from warfare, the tradition of the female warrior continued.[27]

— — —

This is a good point to note that a few of the unidentical twins we are considering here have purely arbitrary boundaries, are separated by what amount to no more than fine lines. Even when the frontier between them may be a river or the height of the land, the barriers are usually easily crossed. Scotland and Ireland, despite their similarities, are an exception, though, because they are separated by a barrier that requires special effort to cross. And yet, while the frequently stormy Irish Sea lies between them, traffic over the waters was and is frequent. Less than a hundred years after Saint Patrick's Christianizing mission to Ireland, one small island in the sea became a veritable stepping stone between them, with consequences that continue today. For a while it was also the centre of preservation of learning in the Western world.

Saint Columba set out to convert pagan Scots and Irish to Christianity from a monastery on the Isle of Iona. Shown here as it was in 1792, it has been restored and is a site of pilgrimage for both Roman Catholics and Protestants. *Source: Archaeologia Scotica volume 1, 1792, public domain.*

Today the Isle of Iona is a place of pilgrimage for many Christians, but when Saint Columba was exiled to it after a bloody battle, it was just an island in the Irish Sea from which Ireland could not be seen.[28] Columba vowed to save as many souls as had died in the conflict, and so in 563 CE he set out to convert pagan Scots to Christianity along with the pagan Irish who had taken up land—colonized, you might say—the wild and empty Scottish coast. In addition, the monastery he set up on Iona became a centre for the copying of manuscripts—not only sacred texts, but also some classical authors and poetry in Gaelic, the first in a vernacular language to be written down.

Monks trained on Iona were sent out elsewhere in Scotland and to the Continent. The monasteries founded by monks trained by Saint Columba (or inspired by his work) have been credited with saving volumes of religious and literary heritage that otherwise would have been lost. In effect, they kept the flame of learning alive during what used to be called the Dark Ages. When Vikings invaded the British Isles, the monastery on Iona was razed, but it has since been rebuilt, becoming a holy place for Roman Catholics and Protestants alike.

Ah yes, the Vikings! They were only one group to covet the islands over the next several centuries. The Angles and Saxons invaded what would become England, too, and held sway there for six hundred years. All the while, in Ireland and Scotland various clan leaders fought each other, claiming sovereignty over lands and cattle. "Petty kingdoms" is what they are sometimes called, and sometimes what was at stake indeed seemed petty. Then, near the turn of the millennium, forces began to come together that foreshadow the political reality of today.[29]

One of the most important was the Norman Conquest of England in 1066. The victory did not extend to Scotland, but Norman title to Ireland came in 1171 along with a victory in a dispute over who should be the king of Dublin. Norman-descended aristocracy, who had become the rulers of England, gave lands in the southern part of Ireland to their friends in an attempt to control the island. But that didn't work, because rather than being a fifth column in Ireland, the Anglo-Norman settlers eventually "went native." Indeed, they would later be an irritant in English-Irish relations for hundreds of years since they had title to much land and had their own ideas of what should be done with it.

It wasn't until the start of the seventeenth century that England began a much more ambitious and—from the English point of view—more successful colonial program when it began to set up "plantations" in Ireland. Anyone who is used to seeing that term applied to vast agricultural estates tilled by slave labour—as in the Caribbean, Brazil, and, closer to home, the American South—will find that use of "plantation" more than a little odd. I certainly did, but the truth is that the first English "plantations" were established in the Old World, not the New, and what was being "planted" wasn't agricultural crops but people.[30] They began in the southwest of Ireland under Henry VIII, but a much larger, officially encouraged program started under James I and involved the mass confiscation of land, particularly in the northeast.

At the time, the region, now called Ulster, was the most closely linked to Scotland, the most deeply Gaelic, and potentially the most troublesome for the English. Part of the initiative was to replace the rural population with English-speaking Protestants while punishing Irish troublemakers. "Sturdy English yeoman" would better husband the land than the supposedly impe-cunious Irish, who, according to the English attorney general for Ireland in the seventeenth century, did not "plant any gardens or Orchards, Inclose or improve their lands," so that "all the Irish Counties are found so wasted and desolate at this day."[31] To correct this, the English proposed dividing the land into geometrical parcels instead of the patchwork of fields and pastures that for several hundred years had served a population based on cattle and a few crops. For justification and precedent the English looked to accounts of the Romans scourging Carthage. The Irish, says historian Ben Kiernan, were con-sidered "contemporary counterparts of the barbarians Rome had conquered and enslaved, with no more right to their lands than any surviving inhabitants of the scorched earth of Carthage." Kiernan goes so far as to call the sum of English interventions and actions in Ireland "genocide."[32]

For the English to compare themselves at that point in their history with the conquering Romans was perfectly consistent with their growing aspira-tions. Significantly, the Ulster plantations began in the same decade of the seventeenth century as the first successful English colony in North America, Jamestown, in what is now Virginia. That region would become rich through the other kind of plantation agriculture, but that was in the future.

In Ireland the scene was thus set for nearly three hundred years of conflict, with several armed uprisings that were brutally put down, and with a litany of punitive measures that meant second-class citizenship for the Catholic Irish and economic control in the hands of absentee landlords.

The English made no attempt to establish plantations in Scotland, even though it could be argued that agriculture there was no more advanced than in Ireland. One reason for that was because Scotland had strong kings, and by then its ruling Stewart dynasty had such close ties to the English throne that the king of Scotland would become king of England after Elizabeth I's death in 1603. Another reason had to do with religion, which we'll get to shortly. Suffice to say that Scots were well represented among the colonizers in the Irish plantations. Tradesmen from City of London guilds were encouraged to move too. They were given lands to develop around Derry, which was renamed Londonderry. By the end of the nineteenth century the six counties of the Irish northeast had become a socially segregated society in which Protestants were said to have the best jobs and the most clout, and the minority Roman Catholics were much poorer and resentful of their diminished position. When Ireland finally became an independent nation in 1922, the Ulster counties remained part of the United Kingdom, with long-lasting repercussions, some good, some very bad.[33] In the latter must be counted the Troubles, a thirty-year period of armed conflict between Irish nationalists and Ulster unionists that ended in the Good Friday Accords of 1998. Among the things agreed upon then were institutional arrangements for cross-border cooperation between the governments of Ireland and Northern Ireland, including a border that has become to all intents and purposes an open one—and which played a huge role in the United Kingdom's withdrawal from the European Union—but that's getting ahead of the story.[34]

— — —

The night of July 11, 2016, I was working on this chapter, and monitoring what was going on in Northern Ireland at the beginning of a holiday that in the past has seen rioting and even death. What was being celebrated was the victory in 1690 of forces of the Protestant William of Orange over those of the Catholic king of England, James II, at the Battle of the Boyne, which took place at the river of the same name, about 48 kilometres (30 miles) from Dublin. No matter that the conflict took place 326 years previously: feelings still ran high.

But the BBC reported that it was not a "particularly boisterous" celebration in Londonderry and the other cities of Northern Ireland. In all, 42 bonfire-related calls were received on this night when Ulster unionist sympathizers (often called Orangemen) light huge blazes, some set in towers of wooden pallets as high as 10 metres. Nevertheless, two row houses in the Shankhill section of Belfast were gutted when wind sent cinders from a nearby bonfire their way.[35] The next day, July 12, thousands of Orangemen marched in 18 parades commemorating the victory of William of Orange, or "King Billy."

Arlene Foster, Northern Ireland's first minister at the time, took part in the activities along with her children. She told the BBC, "this really is about celebration, it's about doing things in a way that we've done so for generations. I always think of the Twelfth when I used to get two sandwiches in a plastic bag and a bun and that is still going on today but it is good fun and we really enjoy ourselves every year."[36]

I suppose it's nice to learn that only two houses were burned down. Indeed, things have been much nastier in the past. That probably shouldn't be surprising, since the holiday was born of hate at a time when family members were at war with each other. The bare-bones story is that in the Battle of the Boyne, William of Orange was fighting forces led by his father-in-law, James II. At issue was the English throne that his wife Mary claimed, and which he would subsequently share with her.

But the dispute was more than an affair of a dysfunctional family (and it's clear that the British monarchy has had quite a few of those over the centuries.) It had its roots in the religious differences that developed after the winds of change blew through the Christian world beginning in the sixteenth century. The Catholic Church had travelled a long way since the days when Saint Columba spread the Good Word to the Scots. Critics within and without the church were ready to sweep away excess, renew understanding of sacred texts, and clean up corruption. The first salvo is said to have sounded when Martin Luther nailed his ninety-five theses to the Castle Church door in Wittenberg, an action that resounded throughout Western Christendom.[37]

In the British Isles unease with religious doctrine was compounded by the lack of a male heir to the English throne. Henry VIII's first wife, Catherine of Aragon—daughter of Spain's Ferdinand and Isabella—gave birth to six children in nine years, but only one, Mary, lived more than a few days. The official story is that Henry wanted a son in order to make sure his succession would go smoothly, and to avoid the kind of conflict that had accompanied so many royal ascensions to the throne in the past (and would in the future, as witness the Battle of Boyne.)[38]

The unofficial story was that Henry had his eye on luscious Anne Boleyn, and thought she was far more interesting than Catherine, as well as more likely to give him a healthy son. So he asked the pope for an annulment in order to remarry on the grounds that Catherine had been married briefly to his late older brother Arthur. An ecclesiastical prohibition forbade men marrying their brothers' widows, but at the time that Catherine and Henry married, no one seemed too concerned, just as Rodrigo Borgia—the future

Pope Alexander vi—fixed things forty years earlier so that Catherine's parents, Ferdinand and Isabella, could marry even though they were cousins. But when Henry made his request, the pope of the day, Clement vii, was much less accommodating than Borgia and refused to annul Catherine and Henry's marriage. (Catherine insisted all her life that her brief marriage to Arthur had never been consummated, which may have weighed in the pope's decision.) Henry was furious, and declared himself beyond the pope's power. The result in the end was the Church of England, with Henry as the head.

The Scots were lukewarm to the idea of Henry's breakaway church, and in the Highlands it never was well accepted: a significant percentage of people remained Roman Catholic. Furthermore, the many other Scots who found fault with the Catholic Church wanted something quite different from a church set up mainly to solve a royal problem that was as much dynastic and libidinous as theological. Many found the rigorous ideas of John Knox, a disciple of the Swiss Jean Calvin, more appealing. The upshot was that within fifty years the Kirk of Scotland was a going concern. ("Kirk" is the Scottish and Northern English term for "church.")[39]

And therein, quite possibly, lies the motor propelling Scotland's amazing economic success in the eighteenth and nineteenth centuries. The key was what otherwise might be considered a small part of the kirk's program. John Knox and his followers held that every person had a right—nay, an obligation—to study God's word. To this end everyone, girls as well as boys, must learn to read. Therefore a law establishing a system of parish schools was passed by the Scottish Parliament in 1696, the first of its kind in Europe. The resulting rise in literacy gave Scotland a massive lead in the major economic and industrial transformations that were just around the corner.[40] Literacy would have a striking effect centuries later when Tunisia put millions into educating its people, while in the late twentieth century Ireland would become the "Celtic Tiger" in part by playing catch-up and emphasizing education at all levels. But back then the consequences of this commitment to literacy were unforeseen except for the salutary effects on one's soul that were supposed to come with a better knowledge and understanding of scripture.

The story was quite different in Ireland. Even though Henry viii declared himself king of Ireland in 1541, ten years after he began to separate the Church of England from the authority of the pope and three years after his excommunication from the Roman Church, Ireland remained solidly Catholic except for the counties that are now Ulster and around Dublin. As the Reformation gathered steam it became very uncommon for a European country or province to have a religious allegiance other than that of its ruler. There's a Latin term for the concept—*cuius regio, eius religio*—which means "whose realm, his religion," and which developed in the Holy Roman Empire as the kings of several countries parted ways with the pope during the sixteenth century.

But in Ireland, as historian John Scally writes, "a lively movement of reform" had already begun among the Augustinians, the Dominicans, and the Franciscans, who were allied with "families in Gaelic society who were in charge of maintaining and cultivating the native culture. The fact that their religion had been rooted within the culture before the reformers arrived would indeed make it very hard to uproot and replace it with a new form of religion." Furthermore, Scally writes, "tensions began mounting between the Old English (local elites) and English rule in Ireland, resulting in their loyalty tilting away from the English monarchy." In effect, the Irish were the "first to find faith and the last to lose it."[41]

— — —

By the end of the seventeenth century, the common folk in both Ireland and Scotland were desperately poor. Scotland was in the Lean Years, the seven years of "sunless, drenching summers...bitter frosts and late snow." Tens of thousands of people died, and it was, says historian Arthur Herman, "a benchmark of collective misery and misfortune."[42] In Ireland, the Gaelic-speaking, predominately Catholic population was being moved off the land they'd lived on for generations and onto marginal territory, where they lived in poverty and helplessness. They would briefly enjoy better conditions after the introduction of the potato at the end of the eighteenth century, but when a blight was unintentionally imported in the 1840s, more than a million died, and another million immigrated.[43] Even many of the largely Protestant people "planted" in Northern Ireland and who were generally better off than the Catholic Irish left during those bad times: between 1717 and 1776, some 250,000 people from Ulster came to North America, 100,000 as indentured servants.[44]

Nevertheless, by the end of the 1700s Scotland had a literate workforce that was both productive and innovative, so when the Industrial Revolution began the Scots were ready to take advantage of employment possibilities. Lowland Scots were tempted off the land to work in the booming cotton mills, iron foundries, and dye works. They were joined by some Highland Scots forced off their hills by the Highland Clearances,[45] and also by Irish who crossed the

Irish Sea in search of work: by 1830 one out of five in Glasgow's workforce had been born in Ireland.[46]

Increased opportunity in Scotland and England was not enough to provide for an increasing population for very long, though. To be sure, there appears to have been an adequate food supply due to the introduction of new foods like maize and, until the Potato Famine, the potato. Certainly infant mortality decreased dramatically: between 1726 and 1751, the death rate in the first year of life in the British Isles was 195 per 1,000, but between 1821 and 1826 it had dropped to 144 per 1,000.[47] But babies grow up and need jobs, food, and places to live. Population pressure increased as a consequence, and more immigration followed. Between 1821 and 1911, 9.6 million residents of the British Isles left for the United States and 3.4 million for Canada. They were encouraged to believe that there was plenty of room for them in the New World, although that space had been gained by a colonial genocide of colossal proportions, which we'll talk about more in the chapters about New Hampshire and Vermont and about Alberta and Saskatchewan.

My grandfather's father, Lachlan McGowan, was one of these migrants. Born near Glasgow in 1837, he shows up in Port Hope, Ontario, in August 1858 marrying Hannah Clifford. His son, my paternal grandfather, David Donald McGowan or McGoun (he spelled it both ways), was born in 1876 in Napanee, Ontario, but didn't stay long. At the age of sixteen he apparently "lit out for the territory," as Huck Finn would say: after his birth, he next enters the official record in the form of the 1901 Territorial Census for Stirling, Assiniboia West, in what is now Alberta. Why he left Ontario is shrouded in mystery. Probably he got in trouble of some sort, like those misfits who were sent out by the Carthaginians.

My other grandfather, John Frederick McDonald—the name he went by for most of his life—came from a family that had been in North America longer. He was born Frederick MacDonald into the MacDonalds of Sherbrooke, Guysborough County, Nova Scotia, a family that settled there at the end of the eighteenth century. But after a detour through Massachusetts, where he added "John" to his name, he headed west to Canada's newly organized province of Saskatchewan. The family story is that he moved to profit from the pure air of the Prairies because he had tuberculosis, but why he left Nova Scotia to begin with is as much a mystery as my other grandfather's flight from Ontario. Certainly in later years, after his lungs were clear and he'd crossed back over the border and settled with his family in the United States, he completely erased that part of his life from his biography, claiming that he and his wife Mary Belle, also born in Nova Scotia, were "native-born" Americans.

Whatever their dramas, they profited from hundreds of years of colonialism, which meant that there appeared to be room on this vast continent for them to reinvent themselves. So did the several million Irish who, while they may have been victims of British rule at home and were discriminated against after their arrival in the United States and Canada, also profited from opportunities in the Americas.

Today Ireland and Scotland are political entities both different and similar. Ireland, after the partition of the North, became an independent nation in 1922, and formally the Republic of Ireland in 1949. Scotland has won more and more autonomy, but it remains in the United Kingdom: voters turned down independence in a referendum in 2014. Many Irish didn't understand how the Scottish could have refused the offer to become an independent nation without fuss, when they had to go through decades of conflict and civil war. The difference in attitude toward the United Kingdom is something that clearly distinguishes the two states from each other.

But there are other things that both Ireland and Scotland have in common now, including a concerted attempt to keep alive or resuscitate their respective forms of Gaelic. Irish Gaelic has been the national and official first language in Ireland since independence, while in 2005 Scotland formulated an official plan aimed at "securing the status of the Gaelic language as an official language of Scotland commanding equal respect to the English language."[48] How successful these efforts have been is in question—despite the requirement that Irish Gaelic be taught in all state-financed Irish schools, the number of Gaelic speakers in that country has stagnated—yet obviously interest in the old tongues is something that persists in both polities.[49]

Then there's that other thing that once united, and has since divided, Ireland and Scotland: religion. As noted, antagonism between Catholics and Protestants has been endemic in Ireland and Scotland since the Reformation. It continues: google "sectarianism Scotland Ireland" and you'll get hundreds of thousands of articles, some soul-searching, some inflammatory. However, religion is playing a decreasing role in both countries. Ireland, so strongly Roman Catholic for centuries, has become much less pious. Between 1972 and 2011, weekly church attendance by Irish Roman Catholics fell from nearly everybody (91 percent) to less than one in three (30 percent).[50] This secularization has influenced political decisions.[51] The attitude toward both homosexuality and abortion has radically changed as well. The former was illegal until 1993, but in 2015 Ireland became the first country to approve marriage between people of the same sex by referendum.[52] What's more, a gay man—the son of an Indian immigrant no less—became prime minister in 2017.[53] Then in May 2018 the Irish voted by a ratio of nearly three to one to allow abortion.[54]

Scotland today is the least religious part of the United Kingdom, with only slightly more than half of the population saying they were Christian in 2011, a decrease of more than 10 percent since 2001. More than a third (37 percent) said that they had no religion, an increase of nine percentage points.[55]

Another point in common is relatively good economic conditions. As late as 1991, when the cult film *The Commitments* was made, it wasn't completely off the wall to suggest that a bunch of young white kids from Dublin could convincingly make soul music because their circumstances were as bleak as that of African-Americans. Unemployment was 19 percent that year,[56] and for much of the century to that point Ireland had ranked near the bottom of European countries when it came to educational attainment. But in 1967 free education for all was inaugurated, setting the stage for Ireland to become the Celtic Tiger of the late twentieth century.

Thirty years after the start of major investment in education, Ireland began to reap its profits, writes commentator John Fitzgerald. "It was only in the 1990s that the rising skills of an educated workforce, combined with an open economy and EU membership, enabled Irish living standards to catch up to our European neighbours." He adds, though: "Wiser policies should have delivered this outcome decades earlier."[57] Be that as it may, audacious economic policy at the end of the twentieth century also helped boost Ireland's economy, with unemployment dropping to a record low of 3.9 percent in 2000.[58] The economy stumbled badly following the collapse of the Irish property market in 2008 and the onset of the Great Recession, but by December 2018 the unemployment rate had been back down to 5.3 percent for months.[59]

At the same time, Scotland reported 3.7 percent unemployment, its lowest unemployment rate on record.[60] Not too shabby, one might say, but some observers saw clouds on the horizon, particularly as the spectre of the United Kingdom leaving the European Union became more and more real. Scotland's business minister, Jamie Hepburn, told the BBC: "Brexit remains the biggest threat to Scotland's prosperity." Those fears lay behind Scottish voters' rejection of Brexit in the 2016 referendum, as they did in Northern Ireland. Scotland's constitutional relations secretary, Michael Russell, in a letter addressed to all members of the British Parliament in January 2019, wrote that Brexit "will make us poorer, diminish our rights and damage opportunities for future generations."[61]

Ireland, of course, is not directly affected by Brexit, although the fate of the "soft" border between Northern Ireland and the republic was one of the most contentious points in discussions between the United Kingdom and the European Union over the terms of the divorce arrangement. Immediately after the Brexit referendum, some voices were heard calling for Scotland to unite with the entire island of Ireland, a new country that would remain in the European Union, but that seems highly unlikely now. The Scottish Nationalist Party lost ground to the Conservatives in the general election of June 2017, and the results were initially interpreted by SNP leaders as a repudiation of the idea of another referendum on Scottish independence.[62]

Since the 2017 election the Conservatives in Westminster have ruled with a minority government, propped up by Arlene Foster—she who had been first minister of Northern Ireland and liked the July 12th celebration so much— and nine other members of the Democratic Union Party. They were enticed into supporting Conservative prime minister Theresa May by the promise of £1 billion to be poured into Northern Ireland over the next few years.[63] Since then, Foster and the DUP have been key figures in the Brexit drama, and when May faced a no-confidence motion in January 2019, they gave her essential support: the House of Commons split 325 to 306, but had the DUP voted the other way, the score would have been 316 to 315—far too close for comfort for a government in the midst of trying to negotiate with Europe.[64]

What strange, convoluted politics, almost as convoluted as the gold, silver, or bronze wires that the Celts twisted together to make their signature torcs.

VERMONT AND
NEW HAMPSHIRE

The change was obvious as soon as we crossed the Connecticut River. We'd been driving all day after leaving Montreal early on a glorious October morning of the Columbus Day/Canadian Thanksgiving weekend, a couple of years after we'd arrived in Canada. A friend who had worked in both New Hampshire and Vermont had offered to give us a personalized guided tour, and we jumped at the chance to see the real New England. Little did I know that the outing would be so important in my reflections about what makes certain places simultaneously alike and different. Until then I'd always considered New England more or less a single entity. But I was wrong. A very direct line extends from that mini–road trip to this book, which is ending, I believe, in some reflections on how places change and how to effect change.

Our route took us south from Montreal along the east side of the Richelieu River, through rich farmland as flat as the San Joaquin Valley in California, where my husband grew up. Then we came to the long valley of Lake Champlain on our right, with the Adirondacks on the other side of the lake, and the Green Mountains off to our left. At St. Albans we turned east and south, driving through green, rolling hills then climbing into the uplands. Beautiful, beautiful: farmland and rounded, purple peaks, green fields, and blazing leaves. Villages that looked straight out of Currier and Ives etchings. Church steeples, white fences, silence except for the wind whenever we stopped the car to take a picture. Even when we met Interstate 89 and turned south we ran through picturesque country. At places we could see across the Connecticut River into New Hampshire, where the landscape didn't look that much different.

But late in the afternoon, when shadows had already obscured the valleys, we met up with Interstate 91, coming down from the Canadian border by another route. That's when we crossed the river and entered New Hampshire, and the difference between the states, so similar in many ways, became apparent. One of the first things that met our eyes was a big billboard announcing that New Hampshire had no sales tax.

No sales tax! My initial thought was that it was a little strange, but not something game-changing. Not true, our friend said. This was a very big deal, indeed, because Vermont had recently started to charge one, while New Hampshire was bound and determined not to. Taxation was—and is—a dirty word in New Hampshire: it still has no sales tax. But by the end of the 1960s, legislators and a sizeable part of the population in Vermont had begun to recognize that taxation is what we pay for civilized society. The other remarkable thing was the sign itself: long before billboards were forbidden on US interstates, Vermont had cracked down on roadway signs; scenic beauty, a large portion of Vermonters had agreed, was good business in a state where tourism was becoming increasingly important economically. This New Hampshire sign was very likely in contravention of nationwide highway rules, but obviously people who had "Live Free or Die" as a state motto weren't about to knuckle under to federal bureaucrats without a fight. Decades later that hasn't changed in many important respects.

Today Vermont and New Hampshire still have a great deal in common. Vermont's population is somewhat less than half the size of New Hampshire's—623,657 compared to 1,350,575, as of 2017—and has a bit more area—24,906 square kilometres (9,216.66 square miles) compared to 24,214 square kilometres (8,952.65 square miles).[1] But—and in the American context, where race and age correlate closely with political ideas, this is extremely important—the percentage of both populations that is white is almost the same—95.2 percent in Vermont versus 94.2 percent in New Hampshire—while a good chunk of the population is over sixty-five: 16.4 percent in Vermont and 15.4 percent in New Hampshire.

New Hampshire has a higher per capita income than Vermont—$36,914 versus $31,917 in 2017—compared to the national figure of $31,177.[2] In both states about 71 percent of the population own their own homes, a characteristic that usually means stability and more conservative values.[3] The comparable figure for the United States as a whole is 63.8 percent.

Politically, both states have been very much alike for long stretches of their history. Between them they have produced three of the most boring presidents in US history: Chester Arthur (1881 to 1885) and Calvin Coolidge (1923 to 1929) were born in Vermont, while Franklin Pierce (1853 to 1857) was born in New Hampshire.

Vermont voted for the Republican candidate in every presidential election from 1856 (the first after the party was formed) until 1960, a record among US states. New Hampshire's loyalty to Republicans has been almost as great: in the same period it voted all but five times for the GOP candidate. (The exceptions were Franklin Delano Roosevelt in 1936, 1940, and 1944, and Woodrow Wilson in 1912 and 1916.[4]) Both states went for Lyndon Johnson in 1964, but went Republican again until 1992, when they both voted for Bill Clinton. Since then, Vermont has remained solidly Democrat, while New Hampshire voted for George W. Bush in 2000. In the contentious election of 2016, Vermont went decisively for Hillary Clinton, but on Election Night in New Hampshire the tally swung back and forth between her and Donald Trump until all precincts reported. The final results showed Ms. Clinton with 47.6 percent of the vote (348,521 votes) versus Trump's 47.2 percent (345,789 votes).[5]

— — —

The differences between the two states go back much further than recent elections, however. On the map they look almost like two wonky right triangles you might get if you tried rather drunkenly to bisect a rectangle. Vermont's the one to the west with its base to the north, while on the other side of the Connecticut River, which runs more or less diagonally across the rectangle, New Hampshire has its base to the south. The river runs roughly between the two arcs of mountains for which the states are famous: the Green Mountains in Vermont and the White Mountains in New Hampshire.[6]

This geology is the basis for a crucial difference between the two states. Once upon a time, back before the first plants and animals began to colonize dry land, Vermont was under water at the edge of an ancient continent where sediments slowly collected, and New Hampshire didn't exist except as some islands out to sea. Then came two waves of mountain building. The first formed the Taconic Mountains in southwest Vermont and parts of New York and Massachusetts. The second came about 375 to 335 million years ago when two plates of the earth's crust collided. The result was the Appalachians, the vast chain of mountain ranges that extends from Kentucky all the way to Newfoundland. The Green Mountains and the White Mountains make up two of the most scenic sections.

But mountains created in the same long period of colossal geologic unrest aren't necessarily the same. The Green Mountains are composed in large part of those sediments laid down in shallow seas, only folded and transformed in the mountain-building phase. The White Mountains, on the other hand, are mostly melted and crystalized rocks produced during what geologists Nancy Bazilcuik and Rick Strimbeck call the "geological train wreck"[7] that saw two sections of the earth's surface pushed together with immense energy. The former mountains are generally made of softer rocks and the latter, more resistant ones. The result—even after thousands of years of being scoured by glaciers—is that the White Mountains are craggier and comparatively ill-suited for agriculture. The range includes sixty-six mountains higher than 1,200 metres (3,937 feet), compared to only nine in the Green Mountains:

The White Mountains of New Hampshire are far more rugged than the Green Mountains of Vermont, a fact that has influenced the two states' histories. *Source: "Crawford Notch" by Thomas Cole, 1839. Public domain.*

Mount Washington, the highest, is 1,917 metres, or 6,280 feet. Samuel de Champlain gave the White Mountains their name, so the story goes, not from snow fields—unlike the highest peaks of the Rockies and the Cascades, in western North America, they are not snow covered all year—but from the shining white granite of the peaks that he spied from out at sea when he charted the East Coast.[8] (He also named the Green Mountains—*les verts montagnes*—which became, of course, the state's name, Vermont.)

Like Scotland, Ireland, and Doggerland, the region was uninhabitable until the retreat of the last Ice Age. Slowly, as the glaciers melted, plants and animals invaded from the south, and by about 7000 BCE bands of Indigenous people had begun to live in the land, now covered by extensive maple, beech, and birch forests.[9] When the first Europeans arrived—Champlain ventured south from New France (now Quebec) along the Richelieu River and into Lake Champlain in 1609—tall white pines nearly two metres in diameter weren't uncommon, and perhaps 95 percent of Vermont was covered by forest.[10]

New Hampshire also was covered with big trees. In the charter governing the colony, the English Crown claimed all trees with a diameter of 24 inches at a height of 12 inches from the ground for "better providing and furnishing the Masts for our Royal Navy."[11] Unauthorized felling could be punished by a fine of £100, the equivalent of about £8,700 today (or between $10,000 and $15,000, depending on whether you're counting in American or Canadian dollars).

The Indigenous people hunted and trapped in the mountains, although it appears their more permanent settlements were along the river valleys like that of the Connecticut and on the shores of Lake Champlain, where they grew corn, beans, and squash (the "Three Sisters") and other crops. Certainly they had a trail system running through the woods, but travel by canoe was more efficient and faster. The predominate culture appears to have been what is usually called Abenaki,[12] which had links to Indigenous communities from Nova Scotia along the southern banks of the Saint Lawrence and into Maine and northern New York, as well as New Hampshire and Vermont. They warred with the Iroquois and Algonquin over territory for several hundred years. But all Indigenous people in the region were badly hurt by a series of epidemics following first contact with Europeans. It's estimated that sixteen separate waves of disease between 1631 and 1758 decimated the Abenaki. This meant that when English adventurers arrived in number at the beginning of the eighteenth century, they found land that appeared to be unclaimed.[13]

Well, it's quite possible that even if the Abenaki had not seen their numbers plummet, the new arrivals would have deemed the land "unused," just as

English argued that the Irish had forfeited their land during the years when "plantations" were being established there. It's no accident, by the way, that the first settlement in New Hampshire was officially called a "plantation" because the colonizers were operating within the same frame of reference. In many respects the story of Vermont and New Hampshire—like that of Saskatchewan and Alberta in the chapter to follow—is the quintessence of what happens when one group of people move into land for which they have little or no legitimate claim.

Americans usually date their earliest permanent settlement to the Plymouth Colony, established by Puritan Dissenters in 1620. (St. Augustine, Florida, dates back to 1565, when the Spanish founded it and named it after Saint Augustine, the bishop of Hippo Regius, whom we met in the chapter on Algeria and Tunisia. But apparently because Florida wasn't acquired by United States from Spain until 1819, it doesn't count.)

Three years after the Pilgrims' landing, an attempt at European settlement was made near Portsmouth, in the territory that would eventually evolve into New Hampshire and Vermont. A royal charter for the province wasn't produced for nearly another sixty years, and there followed several decades of government by the Province of Massachusetts Bay. This period was marked by many conflicts with Indigenous populations, which were subsumed into the bigger fights between England and France. What Canadians and the British call the Seven Years' War, Americans call the French and Indian Wars.

One of the most celebrated raids in that conflict took place at Deerfield, Massachusetts, in 1704 when an Abenaki, Mohawk, Huron, and French force attacked the thriving English settlement. Fifty-six people were killed and 109 were captured and taken north to Canada. Eventually more than half were ransomed and returned to New England. But tellingly, a number of the female captives chose to remain with their captors, apparently because they were so well treated.[14] Yet despite the qualities of the Indigenous society or the strength of its warriors, by the 1740s, one source says, "most of the native population had either been killed or driven out."[15]

The scene was set for New Hampshire becoming New Hampshire and Vermont, Vermont. The communities that had grown up in what's now Connecticut and Massachusetts had succeeded so well that many of them were becoming crowded, and new land was needed to house their people.[16] Unlike other parts of the colonial frontier at the time, the settlers in what came to be known as the New Hampshire Grants—land to the west of Connecticut River and east of Lake Champlain, including all of the Green Mountains—met little resistance from Indigenous people. What did hold them up were English elites in the neighbouring Province of New York who wanted to run this immense territory like another Ireland.

Enter Ethan Allen and his family. He and his six siblings had all been born in Connecticut in the early 1700s, but when they reached adulthood—and all of them did, which was something of a feat in that time—he and his brothers went looking for new land.

It should be noted that Ethan had already been "invited" to leave two communities over arguments. The first time he left a town was in exchange for dismissal of charges of assault. Court records say he stripped "to his naked body and in a threatening manner with his fist lifted up repeated these malicious words three times: 'You lie you dog,' and also did with a loud voice say that he would spill the blood of any that opposed him."[17] The second time, the town meeting in Northhampton, Connecticut, where several members of the Allen family lived, ordered him and his relatives to leave after he became embroiled in religious discussions and scoffed at the author of a collection of sermons. (Note the similarities between the way towns on the frontier and the Phoenicians handled trouble-makers: there was land perceived as vacant and open for settlement, so if you didn't get along, then you knew where to go.)

The land in the New Hampshire Grants looked promising, although at first there appeared not much chance that the Allens and others like them would be able to buy a homestead. Conflict between the powers that be in the Province of New York and those of New Hampshire over title to that land raised obstacles. The New Yorkers hoped to set up a tenant farm system where, as in Ireland, landowners would remain landowners, and those who farmed the land would be peasants. But the governor of New Hampshire had another reading of the province's charter and saw no reason why he shouldn't sell title to land, including the excellent farm land in the Champlain Valley claimed by New York's governors. Lawsuits were undertaken, forceful evictions attempted, and a price put on Allen's head, but nothing was going to force him, his family, and their friends who had bought claims to retreat into the Green Mountains and off the good land.

While there's no doubt that Allen was an irascible character, he also was a born leader, and a well-read one. His father's plan for him had been that he would go to Yale, but just when he'd finished his preparatory tutoring, his father died and Ethan had to take over the family farm in order to support his younger brothers and sisters. When eviction notices came, Ethan was able to galvanize those who had bought land in the New Hampshire Grants into a fighting force in a few short weeks: it helped that the agitation that would shortly lead to the American Revolution was building up. Allen led his Green Mountain Boys (Colonel Benedict Arnold, who would eventually turn traitor to the revolutionary cause, came along too) in a skirmish that resulted in the capture of the British Fort Ticonderoga on Lake Champlain, the first victory in the Revolutionary War. He was less successful in a bid to take Montreal, the first of several attempts by Americans to gain control of what would become Canada (but more about that later). Captured by the British, Allen spent the rest of the Revolutionary War in one military prison after another.

While he was gone, the Continental Congress (the predecessor of the United States government) refused to recognize Vermont as a separate political entity because of New York's and New Hampshire's competing claims to the territory. Allen could not help but be proud when, on his return from imprisonment, he discovered that Vermonters had got around the problem by declaring Vermont independent of all British power in 1777, a year after the adoption of the Declaration of Independence. The constitution adopted by the fledgling Vermont Republic was extremely progressive, doing two things that the American Constitution did not: outlawing slavery and extending the right to vote to all men over twenty-one, whether or not they owned property. This was in stark contrast to the situation across the New York border, in Dutchess and Albany Counties, where only about 10 percent of men could vote: in that realm of tenant farmers you had to be a landowner to vote.[18]

Eventually in 1791 Vermont entered the United States as the fourteenth state, but it continued its independent ways. In 1836 it elected Alexander Twilight to the state legislature, the first black man in the United States to serve in such a capacity. He was also the first African-American to earn a baccalaureate degree from an American college or university, graduating from Vermont's Middlebury College in 1823.[19] In addition to being staunchly abolitionist, in 1880 Vermont also broke ground when it gave women the right to vote and hold office in school districts.[20]

New Hampshire also elected abolitionist politicians, but in some cases it took a lot of work. Particularly notable was John P. Hale, who was turfed out of the Democratic Party at one point for his anti-slavery views, but who was eventually elected to the US Senate from New Hampshire on an abolitionist ticket. In 1846, nearly two decades before the American Civil War, the popular nineteenth-century poet John Greenleaf Whittier wrote a poem in praise of Hale and his state that became a call to arms in the fight against slavery:

Courage, then, Northern hearts! Be firm, be true;
What one brave State hath done, can ye not also do?[21]

Hale ended his political career as the US minister to Spain, appointed by Abraham Lincoln, the quintessential anti-slavery president. It should be noted that the Republican Party for which both New Hampshire and Vermont voted for so long was initially the party of Lincoln, and the Emancipation Proclamation. But that progressive strain was rooted out of the GOP early in the twentieth century.

By then New Hampshire had become an industrialized state. As early as 1831 it had the largest woolen mill in the United States as well as thirty-two major cotton mills. All were powered by the rushing water of the state's rivers, the Merrimack, Salmon Falls, and the Connecticut in particular.[22] By the end of the nineteenth century, New Hampshire had become one of the most urban of the United States, attracting workers to its mills from the farms of New England as well as from Quebec and abroad.

Vermont, on the other hand, remained rural, in part because it didn't have the hydro-power resources of its neighbour. The Connecticut River belongs to New Hampshire as far west as the low-water line in Vermont, which means that the best places to build hydro-power installations are on the New Hampshire side. But also Vermont farmers turned to producing wool for its neighbour's mills. Merino sheep introduced from Spain in the early 1800s thrived on Vermont's hilltop pasturelands. Their wool was easier to transport than cash food crops and fetched better prices, so by 1836 there were as many as a million sheep in Vermont grazing on previously forested land.[23]

The mills in New Hampshire had another advantage that those in Vermont lacked: the goods they made could be shipped from a seaport in the state, while anything made in Vermont had to go overland, which was costly and inefficient even after railroads began to be built in the middle of the nineteenth century. In the twentieth century, Vermont's lack of a seacoast had long-lasting effects on the two states' political development too, according to commentator Chuck Wooster.[24]

To explain, let's return briefly to our sightseeing trip to the two states so long ago. We were surprised that New Hampshire did not have a sales tax then. Alaska, Delaware, Montana, and Oregon did not—and still do not—either, but all of them have substantial other sources of revenue. In Alaska, taxes and royalties on oil production fill the state's coffers. Delaware levies corporate taxes, while in Montana and Oregon income taxes make significant contributions to state finances.[25] But New Hampshire doesn't have oil or corporate cash cows, nor does it charge an income tax except on dividends and interest.

Why? Vermont observer Wooster puts it down to a balancing game that the writers of New Hampshire's constitution played between the coast and the interior, the urban and the rural.[26] Back in the 1920s and '30s both New Hampshire and Vermont considered imposing an income tax, which required the approval of voters. In both states farmers and residents of small towns voted strongly in favour of the measure, but in New Hampshire a 60 percent Yes vote throughout the state wasn't enough for passage. The state's constitution, adopted about the same time as Vermont's in the 1770s, was designed with an eye to the interests that had developed already over its nearly one hundred and fifty years of settlement history. Wealthy merchants on the coast were thought to merit more political clout than the growing numbers of homestead farmers in the interior, so the constitution's writers, Wooster explains, required a two-thirds majority for any new legislation to pass. In his novel *Northwest Passage*, Kenneth Roberts describes urban life in New Hampshire's Portsmouth in the 1770s:

> Every night the lively grandees of Portsmouth make a promenade of Buck Street, gabbling, and laughing, their sticks and high-heeled shoes clicking on the pavement, their silks rustling, their powdered hair and shoe buckles gleaming in the light from tavern doors and shop windows.
>
> Merchants, shipbuilders, lawyers and bankers are there with their wives and daughters; naval officers from the King's ships; army officers from Fort William and Mary in scarlet and gold, their swords clutched tight beneath their arms; members of the King's Council, long crimson-lined coats swinging from their shoulders. All of Portsmouth, as the saying goes...stroll up and down Buck Street in seeming amity on every pleasant evening.[27]

Compare that with Vermont, which in the 1770s had had only two decades of settlement history and an economy based entirely on homestead farmers. No town in Vermont came anywhere near the glories of Portsmouth and as a consequence, Wooster says, Vermont was much more egalitarian—remember the Green Mountain Boys?—so a simple majority was set as the bar for any referendum.

— — —

The income tax question was a turning point in the political history of the two states. Today, certainly, the absence of either income or sales tax in New Hampshire has profound effects. People who hate taxes are much more likely to choose to live in New Hampshire while those who are more interested in what taxes pay for will opt for Vermont, according to historical sociologist Jason Kaufman.[28] Working with Matthew Kaliner, in 2007 he tried to tease out the reasons for the political differences between the states. Kaufman and Kaliner suggested that much of the difference has to do with self-selection, a sort of "idiocultural migration" that suggests that interstate migrants move for cultural reasons, not just economic ones. Any number of comments on Internet message boards these days corroborates the hypothesis: moving to New Hampshire is a way to get away from "Taxachusetts," one posting suggests, and certainly most of the state's population growth has recently come from people who work in the Greater Boston area. On the other hand those who like Vermont laud its "reasonably progressive tax system," its beauty, and its "less materialistic" lifestyle.[29]

Since about the only source of revenue New Hampshire has is its property taxes, they are the third-highest in the United States. Even then they don't pay for services that are taken for granted elsewhere: New Hampshire only instituted free kindergarten for all children in 2007—the last state to do so.

Arnie Arnesen, a New Hampshire radio commentator and former Democratic gubernatorial candidate, tried to explain the difference between the states to CNN viewers in a background piece before the 2016 Democratic primary in which Vermont's Bernie Sanders was up against Hillary Clinton. Both states are largely rural, she said, but New Hampshire is "rural-industrial" and Vermont is "rural-agricultural." "When you're the former," she went on, "what matters is who owns the plant, the mill. Everything is concentrated in the little fiefdom so [New Hampshire] grew up with this difference in orientation. But when you're rural-agricultural there is no fiefdom, the border of the town has no meaning...a different sense of community is created....New Hampshire is all about me, myself, and I, and my town. But in Vermont it's about the state."[30]

— — —

We should be clear about this, however: the attitude toward government in Vermont isn't all left-wing, airy-fairy, eco-friendly hype. Vermonters learned first-hand of the necessity for statewide government action as long ago as 1927, when disastrous floods left 84 dead and 1,258 bridges destroyed or severely damaged.[31] That precedent of state intervention set the scene for more in the middle of the twentieth century, when it became clear to many that one of Vermont's most marketable assets was its pastoral, natural beauty. Protecting and promoting that would be the key to the future as agriculture stagnated and the young left the state for greener pastures—pun very much intended.[32]

Tourism had been important to New Hampshire in the nineteenth century, when the rugged peaks of the White Mountains drew well-heeled visitors looking for "sublime" experiences. "The White Mountains...have had more written about them, probably, than any other mountains, the Alps alone excepted," Allen H. Bent wrote in 1911.[33] For decades stagecoaches and, somewhat later, special railroad lines carried pleasure-seekers to well-appointed hotels in the mountains.

But later in the century the picturesque scenery of Vermont became more in vogue; particularly in the southern part of the state, which is within easy travel distance of Boston and New York City, resorts were developed to cater to these tourists. When the automobile arrived, making trips to more remote locations easier, city dwellers began to snap up old farmhouses in Vermont and convert them into summer houses. Developments intended exclusively for vacationers followed, and by the 1950s the state's leaders, its largely rural population, and the new summer people began to comprehend the conflicts inherent in making the state a tourist destination. In short, there was a consensus that too much development and sloppy regulation promised to spoil whatever was lovely and attractive in the state.

There followed a number of measures designed to balance the interests of tourists, those locals who profited from them, and the visitors who had transformed themselves into new Vermonters. These included restrictions on road construction, regulation of signage (no billboards, you'll remember), and long-term development plans. While there was a certain amount of backlash from people who wanted to profit immediately from the tourist influx, ultimately the need to safeguard Vermont's beauty and ecology won the day.

The 1960s and '70s saw various back-to-the-land movements across the United States, and Vermont was home to several. Some of these were ephemeral, but the attraction of places where it was thought life would be slower and more "natural" led to the arrival of many young people who had—and it's no exaggeration to say this—ambitions of making the world a better place.

— — —

To get an idea of what it was like then, let's take another leap back in time to a July day a couple of years after our first foray into New England. Another friend—this one a great fan of country music—learned of an Old Fashioned Fiddle Music Festival in Craftsbury Commons, Vermont. Why not check it out? we thought. So with several other couples we went down. Another eight thousand people had the same idea.[34]

Craftsbury Common, one of three villages in Craftsbury Township, lies on the top of a hill. From the common, which consists of about two acres of grass surrounded by neat white-painted wooden houses, you can glimpse the Green Mountains to the west, surrounded by rolling hills. It is verdant, beautiful country in the summer, but hard land to farm: some fields are still full of rocks after more than two hundred years of cultivation. The local United Church, of which Reverend Arnold Brown was pastor at the time, sits on the west side of the common; the public school sits on the other. Aside from that, the village in 1972 looked much the same as it did twenty or thirty years earlier, judging from archival photos: no gas stations, no hamburger stands, no trailers, no

Craftsbury Common is the archetypal Vermont village surrounded by green fields.
Source: Jessamyn West, public domain.

tourist cabins. The day of the fiddlers contest, people had to hike down the road to buy beer.

The festival was like Woodstock was supposed to be, only for everybody, not just one age group or one lifestyle. Even the show's main attraction had intergenerational appeal. A good forty years separated the performer, Hazel Henderson, and the audience. She played at about 11 p.m., after some of the older folks had gone home. The remaining crowd, composed largely of people under thirty, huddled together amid beer cans, wine bottles, and the remains of picnic suppers. The air was fragrant with the smell of hay—and of that other kind of grass—and the sky was a velvety blue and full of stars.

Then everyone went wild over Hazel. They stamped and clapped and danced. She beamed, tossed her head, and played on, but the clamour grew so loud that Ken Dahlberg, the emcee, had to stop the show.

"Come on, folks," he said. "The judge can't hear Hazel and that's pretty unfair, don't you think? Why don't we all simmer down and give her a chance."

The three or four thousand remaining spectators roared in agreement, and then grew quiet, listening to the seventy-three-year-old grandmother do her stuff.

"I suppose if we don't do anything else we've really accomplished something when all those young people get so excited over an old lady like Hazel," Reverend Brown, who was one of the contest's founders, said afterwards. "It's not very often when the young and the old appreciate each other like they did there."

Well, those young folks are old now, and some of them have made their mark. One of the newcomers to Vermont at the time of the fiddle festival was Bernie Sanders, whose surprising campaign for the 2016 Democratic nomination, though ultimately unsuccessful, inspired thousands across the country. Sanders cut his political teeth in Vermont, running first as an unsuccessful third-party candidate for both governor and US senator on the ticket of the short-lived Liberty Union Party in 1972, the year of the festival.[35] He ran unsuccessfully twice again, but in 1980 he won the non-partisan office of mayor of Burlington (the state's largest city), a position he held for eight years.

Sanders has consistently plumped for government investment in things like housing and education while championing planning when it comes to development. Indeed, his first victories as Burlington mayor were in stopping a plan to redevelop the city's waterfront with high-end housing and office buildings. The result of his leadership is a mixed-use downtown Burlington, which gives the city its current charm.

After his mayoralty, Sanders served for sixteen years as an independent in the US House of Representatives. He stepped down to run for senator in 2006, a post he held up to and during his run for the US presidency. At various times he has described himself as a socialist or a social democrat, even though he caucused in the Senate with Democrats. In December 2015 he joined the Democratic Party officially in order to seek the nomination as the Dems' presidential candidate, but in 2018, when he ran successfully once again for re-election to the US Senate, he turned down the Democratic nomination for senator and ran again as an independent. Party labels obviously didn't matter because he won more than two-thirds of the vote in the November midterms.[36]

Vermonters also split their vote when it came to the races for governor and Congress. The Democratic candidate for Vermont's one seat in the House of Representatives won by a slightly bigger margin than Sanders, but the incumbent governor, Republican Phil Scott, was re-elected with 55.4 percent of the vote.[37] Nevertheless, even this victory is an indicator of how far Vermont has come since the days when it was solidly Republican. Scott is the only Republican elected to a statewide office in Vermont since 2013, and during his first term he distanced himself from President Donald Trump, and came out in favour of restrictions on firearm ownership.[38]

There has been political change in New Hampshire too. In 2018 Democrats took both houses of the huge state legislature—which includes a 400-member House of Representatives and a 24-member Senate—for the first time since 2012. And while the Republican incumbent governor Chris Sununu won re-election, he'll face a House and a Senate with large Democratic majorities. The agenda put forward by New Hampshire Democrats is decidedly to the left: on election night Democrat Steve Shurtleff, who is now speaker of the House, promised "to expand access to affordable health care, and create the paid family and medical leave program." As well he called for putting "an end to the opioid crisis," which has hit New Hampshire particularly hard.[39] It had the second-highest rate of opioid-related overdose deaths in the United States in 2016—35.8 deaths per 100,000 persons. That rate was nearly three times higher than the national rate of 13.3 deaths per 100,000, and nearly twice that of Vermont's (18.4 deaths per 100,000)—a states whose health services have been at the forefront of innovation when it comes to reducing the toll of the opioid crisis.[40] That Scott mentioned the growing problem at all is a sign of increasing recognition that, even in a state famed for its hard-core individuality, government intervention can be necessary.

So, despite its white and rural population, it seems the people of New Hampshire, like those in Vermont, are far from being wedded to the kind of anti-government politics that marked the state's past. What this bodes for politics in other parts of the United States with similar demographics is unclear. We'll explore that more in the chapter on the United States and Canada's status as unidentical twins. Suffice to say for now that New Hampshire and Vermont show that attitudes can change over time. Sometimes that change is part of a cycle of self-selection, proof that population movements can happen for cultural reasons, not just economic ones. And sometimes it comes when people realize where their real interests lie, like their need for access to medical care, including addiction help.

ALBERTA AND
SASKATCHEWAN

Twenty years after our eastward trek to Montreal, we crossed Canada in the other direction. We were a nuclear family by then: two adults, two kids, a ton of camping gear, and five weeks, more or less, to spend on the road. The idea was to check out Winnipeg—we're city folk, after all—and then go on quickly because there was nothing to see between there and Jasper National Park, right? After that, we'd head for the Pacific coast to visit family.

But our travel plans were turned upside down when we spent a Sunday afternoon recuperating from a long travel day by visiting Louis Riel's house on the Red River in Winnipeg. After learning about the rebellion he led in the middle of the nineteenth century, we decided to visit where the last Métis battle had been fought slightly more than a hundred years before our trip.

The drive might have taken us just a day, had we kept to the main route, but in the end it took us three. We were dumbfounded by the beauty of the country, the yellow canola fields, the blue, blue sky, the thunder clouds in the distance, the sense of space. Montana's motto may be Big Sky Country, and the Mongols—also people of grasslands—may have worshipped the spirit of the Great Blue Sky, but the Canadian Prairies are just as awe-inspiring.

I would like to be able to do the landscape justice, but I can't. However, there is a writer who does—Wallace Stegner, who was born in Iowa and is claimed by the Americans, but who spent several formative years on the Canadian Prairies. In the late 1950s he took a trip home to the community of Eastend, Saskatchewan, where his family had lived. He described what he saw on the way:

> The plain spreads southward below the Trans-Canada Highway, an ocean of wind-troubled grass and grain. It has its remembered textures: winter wheat heavily headed, scoured and shadowed as if schools of fish move in it; spring wheat with its young seed-rows as precise as combings in a boy's wet hair; gray-brown summer fallow with the weeds disked under; and grass, the marvelous curly prairie wool tight to the earth's skin straining the wind as the wheat does, but in its own way, secretly....Across its empty miles pours the pushing and shouldering wind, a thing you can tighten into as a trout tightens into fast water. It is a grassy, clean, exciting wind, with the smell of distance in it, and in its search for whatever it is looking for it turns over every wheat blade and head, every pale primrose, even the ground-hugging grass.[1]

It's a wonderful description, I think. Notice, though, that an important detail is missing. Although he describes travelling from Alberta east into Saskatchewan, he doesn't mention the provincial boundary. Nor do I remember what it was like when we crossed it on the Yellowhead Highway headed west from North Battleford. The reason is simple: the landscape at this point rolls on and on with nothing to impede it; the boundary is, literally, just a line drawn on a map. There is no geographic difference between the two provinces to grab your eye, and there was no real reason to put the boundary there, aside from political expediency.[2]

But the two provinces are quite different, representing two poles in many people's minds: one progressive, the other conservative. (A note here about the names of Canadian federal political parties. For much of the twentieth century the right-of-centre party called itself the Progressive Conservative Party, a name that seems an oxymoron but which reflected the political roots of some of its leaders. Early in the twenty-first century the PCs dropped the "progressive" bit and merged with the more right-wing Canadian Alliance to form what is now called the Conservative Party of Canada. The Liberals, who governed Canada for most of the twentieth century, are usually considered a centre or centre-left party. The New Democratic Party—the NDP—has usually been to the left of the Liberals since it was formed in 1961. To further confuse outside observers, the Liberals in Canada use red as their campaign colour and the rightish forces use blue, while the reverse is the case in the United States. The NDP, always distinctive, is orange.)

Yet Alberta and Saskatchewan may be growing closer in the twenty-first century, for reasons that say much about the way people and their polities change.

— — —

But first let's consider two men who some would say embody the opposing aspects of Saskatchewan and Alberta: Tommy Douglas (1904–86) and Stephen Harper (born in 1959). Their names may mean little or nothing to Americans, but to Canadians, just saying them evokes sharp reactions. Neither one is a native of the provinces that became their homes: Douglas was born in Scotland and Harper in Toronto, and therein lies part of the tale.

I heard Tommy Douglas—the son of a foundry worker who became premier of Saskatchewan, a leader of the NDP, the Father of Medicare, and Kiefer Sutherland's grandfather—speak only once, on a cold and rainy November night in 1982. It was at the end of a conference on the future of Canada's Medicare program, at a time when the still relatively young program was under attack, as it frequently has been.

The speakers who went before Douglas had given their view of how Medicare—one of Canada's defining attributes, many think—had developed. A trade union official spoke about establishing one of the first comprehensive health clinics in North America, an actuary and former Quebec politician told how a universal, public hospital and medical insurance scheme was more efficient and equitable, and a physician explained why doctors in one province who'd gone on strike to block its introduction had ultimately come around to agreeing it was a pretty good idea. The speeches were informative, sometimes even interesting.

Then Douglas, still going strong at seventy-eight, stepped up to the podium to "bat cleanup," as one of the conference organizers put it. His speech wasn't new to anyone who'd been following his career.[3] Mostly it was about how the universal health-care system was threatened by extra-billing and the levying of additional charges even though it was set up to provide free health care "for every man, woman and child in the country." He spoke with his right hand raised in the air, his voice strong and commanding, trained as it was to be heard in large crowds in the days before public address systems when he was a Baptist preacher. His presence was as big as his voice, despite the fact that he'd always been a little guy, never much heavier than the 135 pounds he weighed when he won a light heavyweight boxing title at eighteen.[4]

"We can't stand still," he said. "We can either go back or we can go forward. The choice we make today will decide the future of Medicare in Canada." He paused, and then rose up on the tips of his toes to quote not scripture but the English poet William Blake. His voice intensified, full of conviction, of strength, of hope: "We shall not cease from mental strife / Nor shall our

LEFT: Stephen Harper was Canadian prime minister from 2006 to 2015, and has championed conservative policies throughout his political life. *Source: Remy Steinegger, Creative Commons Attribution-Share Alike 2.0 Generic.* RIGHT: Tommy Douglas: son of a foundry worker, former premier of Saskatchewan, former leader of the New Democratic Party, and the Father of Medicare. *Source: Uncredited photo, public domain.*

swords rest in our hands / Till we have built Jerusalem, / In this green and pleasant land."

The crowd took to its feet with cheers, whistles, and applause. If he'd told us to march out into the darkness and bring some light to the world, we would have done it. I'd like to think that some of those present continued to fight the good fight even after Douglas made his final bow and waved for the last time, after they had gone out into the miserable weather.

All right, it's agreed: he was an orator of rare talent. But he was more. He was named the Greatest Canadian in 2004 in a nationwide competition run by the Canadian Broadcasting Corporation, and even though he never was prime minister, his appeal was perhaps most like American president Barack Obama. That is, his political message was one of hope, that positive change was possible, that we could and should make things better.[5] In 1947 his party, then called the Cooperative Commonwealth Federation, introduced hospital insurance in Saskatchewan. Similar programs were adopted shortly afterwards in Alberta and British Columbia, and then in 1957 a national hospitalization insurance program was passed under a Liberal government. It fell to the Progressive Conservative government of John Diefenbaker to put the plan into effect, and then Diefenbaker set up a Royal Commission, which proposed a national health-care program. By then the Liberals had regained control of the government, but clearly Medicare enjoyed widespread approval across all parties: in December 1966, when the vote on the National Medical Care Insurance Act came in front of the House of Commons, it was passed after a vote of 177 to 2.[6]

Note the healthy cross-party, pan-Canadian support for the idea of Medicare. By then the CCF had become the New Democratic Party, and even though it had only twenty-one seats in the House it, like Tommy Douglas, was clearly punching well above its weight when it came to principled, progressive policies. In the eyes of many Canadians today, Saskatchewan, governed for more than three decades in the last half of the twentieth century by the CCF/NDP, has a reputation for being a beacon and showcase for the ideas of the Left, of social democracy even—a place where Bernie Sanders would fit right in.

Stephen Harper is a different kettle of fish—or basket of saskatoon berries and wild roses, to take the metaphor onto the Prairies. Born into a comfortably well-off family and schooled in Ontario, Harper, fifty-five years younger than Douglas, started off as a Liberal, but moved to Alberta shortly after high school, where he took a sharp turn to the right. The precipitating incident appears to have been the National Energy Program that Liberal prime minister

Pierre Trudeau—father of Justin Trudeau—pushed through in the 1980s.[7] Among the program's stated goals was securing adequate oil for the country at prices that might be lower than those fetched abroad. Clearly aimed at maintaining Eastern Canada's industrial dominance, it was wildly unpopular in the oil-producing Western provinces: "Let the Eastern bastards freeze in the dark," quipped a popular bumper sticker. When Progressive Conservative Brian Mulroney became prime minister, westerners and the Right expected him to act quickly to quash the program. When he didn't, Harper joined the fledgling Reform Party, which later morphed into the Canadian Alliance and then became the Conservative Party of Canada. In short, the names changed but most of the players stayed the same, as social and economic conservatives sought to "Unite the Right" by defeating more small-*l* liberal parties. Their aim was to move the centre of power westward from Canada's central provinces, Ontario and Quebec.

Significantly, Harper appears never to have been interested in provincial politics. From the beginning his eyes were focused on what might be called the main event: what happens in Ottawa. He seems to have always thought that the nation's capital was where changes could be made, where there were enemies worth fighting. The big enemy in his eyes was the Liberal Party, whose supporters like to call it the "natural governing party of Canada" because its power base in Ontario and Quebec ensured it won election after election. Indeed, it governed for sixty-nine years of the twentieth century. But probably just as important for Harper's interest in federal government was the fact that for even longer Alberta politics had been dominated by conservative parties, parties full of young men with conservative ideas against whom he'd have to compete for the top spots.

Front and centre in the credo of Harper and his friends is the belief that the best government is the least government. Harper's attitude toward health care is telling and in stark contrast with that of Tommy Douglas: in 1997 he claimed that "the best system means having a system where you have as many tiers as possible and you bring in as many health-care dollars into this country as possible." The subtext was that getting patients to pay was a good thing. Nothing in his behaviour as prime minister from 2006 to 2015 indicated that he'd changed his mind.[8]

What Harper thought government could and should do was encourage business, and when it came to paving the way for new resource development—particularly things like the Alberta oil sands and pipelines to get that oil to market—he was an enthusiastic fixer. By the time he was soundly defeated in the federal election of 2015, he had become as emblematic of the politics of the Right as Douglas was of the Left. Over his nearly ten years as prime minister Harper attempted, frequently successfully, to undo many of the things that Douglas and his breed of progressive thinkers championed.

— — —

So how did Alberta and Saskatchewan, two political entities born with more or less arbitrary boundaries, throw up such different political cultures?

Both became Canadian provinces in 1905, born of the vast territory that had only recently become meccas for settlement. When Europeans arrived in the Canadian West they met fully developed Indigenous societies who had been hunting buffalo, trapping beaver and muskrat, collecting the region's plants, and travelling the plains for several thousand years since the end of the last Ice Age.

Henry Kelsey was the first European to see the vast grasslands that make up most of today's Saskatchewan and Alberta. In 1690—the year that King William won the battle that secured Northern Ireland as a British province— he was sent out from York Factory on Hudson Bay. An employee (or "servant," as the post was called) of the Governor and Honourable Company of Adventurers of England Trading into Hudson's Bay, he was "to call, encourage and invite the remoter Indians to a Trade with us."[9] At the time, the Hudson's Bay Company was twenty years old and had been given monopoly trading privileges and mineral rights to all the lands drained by the rivers flowing into Hudson Bay. This meant that the HBC purported to control about 15 percent of the North American continent, an area that came to be called Rupert's Land.[10] Like Pope Alexander VI, who divided the world between the Spanish and the Portuguese, when the English king Charles II granted the charter to the HBC he had no idea what that would entail. He was only following the model set seventy years before by Queen Elizabeth I when she chartered the Governor and Company of Merchants of London Trading into the East Indies, which would eventually rule a good part of India.

The HBC was interested in procuring furs, particularly beaver pelts that could be made into hats that were enormously popular. But drumming up business was easier said than done. Few of the Indigenous population lived on the bay because the climate was bad and the land swampy. Hunters and trappers who lived farther away had to be convinced that if they brought their furs to the coast they're receive trade goods of value. Kelsey set out with a sample of what they'd receive, but it's clear that he would not have known where to go had he not been helped by Indigenous people who saw an advantage in doing so. As historian Bill Waiser notes, the groups he encountered had not as yet been badly affected by Europeans, their way of life was reasonably in equilibrium, and Kelsey entered their territory as neither an invader nor a conqueror. What he saw was a good, grassy country inhabited by people who had learned to make the best of its many resources.[11]

The English, of course, were not the only ones who coveted the wealth in furs and other riches of the Western Hemisphere. As we've seen in the chapters about Brazil and Spanish-speaking South America and about Haiti and the Dominican Republic, the Spanish and the Portuguese had been adventuring there for nearly two hundred years before Kelsey made his trek inland. Spanish expeditions going north from Mexico had reached the Great Plains of what is now the United States almost a hundred and fifty years earlier. Waiser suggests that the Assiniboine and Cree peoples Kelsey encountered had probably already heard tales of Europeans since information and goods circulated among the many Indigenous societies along trade routes that covered the continent.

The French had been trading in furs from their base in the Saint Lawrence Valley since the early 1600s. They also had poked around the southern Mississippi region, and by 1682 Robert de la Salle had travelled from the one pole of French influence to the other by going from the Saint Lawrence, through the Great Lakes to the headwaters of the Mississippi, and thence to the Gulf of Mexico. When the French learned of the English setting up shop in Hudson Bay, they sent a warship to take York Factory. Skirmishes followed, and the place passed back and forth between French and English hands several times. It wasn't until the Peace of Utrecht in 1713 that the English finally received Rupert's Land definitively, and even then the boundaries weren't defined.

But neither the English nor the French were interested in settling the vast territory. Controlling it, yes, most certainly, but attempting to turn the countryside into farm land, as the French were doing along the Saint Lawrence and the English were doing on the Eastern Seaboard was not on anyone's agenda. Thus, at a time in the mid-eighteenth century when New England settlements were pushing back from the coast, and Ethan Allen's folks wanted to settle the good land in the New Hampshire Grants, there was no European settlement in Rupert's Land outside a few trading posts—and the only Europeans were men.

When you put young men on the edge of civilization and keep them there for years at a time, you shouldn't be surprised that they find solace and release and maybe love with the young women who live around them. Taking a mistress was frowned on by the HBC but impossible to stop, particularly since a woman's help was essential to accomplishing the many tasks associated with preparing pelts, to say nothing of learning how to survive the winters. The HBC employees on the ground recognized this from the beginning, although the governing council in London didn't understand. Kelsey himself ran up against the policy when he returned from his walkabout in 1692: he refused to enter York Factory unless his partner/helpmate/country wife could come along too.[12]

In the end, countless HBC employees formed relationships with local women, frequently marrying them "in the custom of the country"; that is, under local rites that often were conveniently forgotten once the man was transferred back to England or Scotland or Montreal. The rival North West Company, founded in 1779 and trading out of Montreal and into the country south of Rupert's Land, was much more open to the practice. In 1806 the HBC reiterated its opposition to marriage to Indian women with a ruling that Waiser says indirectly underscores the demographics at play. HBC men were flatly forbidden to take Indigenous wives, but somewhat paradoxically an exemption was made for marriage to the daughters of European traders, who were of course the offspring of unions that had been forbidden. Waiser notes that HBC employees who wintered in the country were understandably concerned for the future of their daughters; this measure was intended "to ensure a more secure future" for them, with the added benefit that "the marriages to incoming traders would also bolster company solidarity."[13] The result of these unions was a mixed-race population that grew rapidly. Some of the sons of the traders were sent back to Montreal to study, while others were apprenticed, but many children became part of a new society that would play an important role in Alberta and Saskatchewan, the Métis.

So things went along for decades, with the HBC and the North West Company fighting for control of the fur trade, while overhunting of beaver meant going farther and farther afield to get pelts. No attempt was made to switch the region to another sort of economy until the early 1800s. By then some of the Métis were providing the North West Company with food from their small holdings and from the bison they hunted. Lord Selkirk, a Scottish nobleman, wanted to set up a settlement where Scots who had been pushed off their lands during the Highland Clearances could grow food to provision the HBC.[14] His choice for settlement was the fertile land near the meeting of the Red and Assiniboine Rivers, now the centre of Winnipeg. The community became home to a mixed-race population that included both Catholic Francophone and Protestant Anglophone families. Most were descendants of English, Scottish, and French voyageurs and traders who had settled down with Indigenous wives. After considerable *Sturm und Drang*, the colony eventually became the centre of Manitoba, and as such might not seem to belong in the story of Saskatchewan and Alberta. But many of the Métis believed themselves badly done by the arrangement, and moved west into what would become Saskatchewan.

— — —

Let's back up a minute to consider who actually was running the show on the Prairies back then. It wasn't a government or a king, but something that former prime minister Stephen Harper would appreciate: a business, the Hudson's Bay Company.[15] Yet no one knew just what were the boundaries of the territory claimed by the HBC (which combined with the North West Company in 1820) any more than the United States knew just how much land it got in 1803 when it paid France $15 million for Louisiana, the vast territory west of the Mississippi River. As it happened, the two territorial claims overlapped, requiring negotiations and some skirmishes. The Louisiana Purchase doubled the territory of the United States, but it took the Lewis and Clark expedition in 1804 to get a better sense of its boundaries.

The Canada we know today was formed in 1867, bringing together what are now the provinces of Nova Scotia, New Brunswick, Quebec, and Ontario. British Columbia, which had been a British colonial outpost, was added in 1871 on the promise of building a truly transcontinental railway: the so-called Intercolonial Railway began in Nova Scotia and ended in Quebec. (Prince Edward Island joined in 1873, but it is not a part of this story.)

By then Canada had gained control of the Prairies when it acquired the foundering HBC by purchasing Rupert's Land from it. The company had become considerably less profitable largely because the fur-bearing animals that were its stock in trade had been hunted to near extinction. What was needed, the company's friends believed, was a government bailout on a grand scale. When word went out that the HBC wanted to unload its territorial interests, the United States, which had just bought Alaska for $7.5 million from Russia, was very interested. But under pressure from Great Britain the HBC agreed in 1869 to sell its rights to Canada for $1.5 million. Of course, little thought was given to the people who lived there—the Indigenous population, the Métis, the former HBC employees, and other settlers. As Prime Minister John A. Macdonald wrote to his Francophone political ally George-Étienne Cartier: "All these poor people know is that Canada has bought the country from the Hudson's Bay Company and that they are handed over like a flock of sheep to us."[16]

That produced protests, particularly over title to the land farmed by Métis. As I said earlier, the trip my family took across Saskatchewan began as a trek to see the scene of the last battle in the North-West Rebellion, at Batoche on the South Saskatchewan River. That 1885 conflict, which ended with Métis leader Louis Riel hanged for treason, has become one of the emblematic events in the history of the Prairies.

Twenty years after Riel's martyrdom (as some would have it) and the completion of the transcontinental Canadian Pacific Railway, another railroad—the Canadian National—was under construction to the north of the CP line. New cities were "strung out along the railroad like beads on a string," one observer said.[17] The newly accessible farm land was booming, immigration from Europe was encouraged, times looked good. The moment had come to create new provinces to govern the burgeoning territory.[18]

Several possibilities of how to do this were proposed, including setting up one big province to be called Buffalo, or two provinces divided along east-west longitudinal lines, or two provinces divided by a north-south line, or even four provinces. That so many possibilities were considered shows that there was no compelling geographical rationale for setting up borders for the new governmental entities. In the end, the dividing line was rather arbitrarily set at the 110th west meridian, and in 1905 Alberta and Saskatchewan were born. (In the United States, settlement went rather more quickly. There the 100th meridian was considered the point where the Great Plains and the frontier began, a frontier which the US Census declared closed in 1890.[19] Oregon, the end of the Oregon Trail, entered the Union in 1859, while the Western states of Washington and Montana were judged sufficiently settled to become states in 1889. Idaho followed in 1890.)

Settlement in Canada speeded up considerably in the first decade of the twentieth century. Between 1901 and 1911 the population of Alberta grew from 73,000 to 374,000, while that of Saskatchewan went from 91,000 to 492,000—in both cases an increase of more than 500 percent. It was not an easy life, though. As one popular song of the period describes conditions in Palliser's Triangle in southeast Alberta and southwest Saskatchewan:

Hurrah for the Palliser, land of the free
Land of the wheat rust, grasshopper and flea
I'll sing and I'll praise it, I'll tell of its fame
While starving to death on my government claim.[20]

People flocked there nevertheless. Maps prepared from 1911 census data show the predominant ethnic groups in both provinces, using colour to indicate which group dominated in a census division: pink is English, blue is "Scotch," green is Irish, grey is French, maroon is Scandinavian, and white is Aboriginal.[21] Pink is the dominant colour on the Alberta map, but Saskatchewan is much more of a patchwork. Among the "English" in Alberta were many Americans who had crossed the line in search of cheaper land, says University of Regina history professor James M. Pitsula, writing in the *Beaver*, the Canadian history magazine.[22] He notes that in the 1920s, US settlers in Alberta outnumbered Brits in rural areas by a ratio of two to one, but the Brits slightly outnumbered Americans in Saskatchewan.

In many respects the governments of the two provinces—both were Liberal, as was the federal government of the day—started out by adopting many of the same measures. Among them was buying out private telephone companies and establishing provincially owned telephone services (Alberta in 1906 and Saskatchewan in 1908). Alberta's was privatized in 1990, but Sasktel still exists as a Crown corporation. However, Pitsula says, the two provincial governments differed early on in that Saskatchewan was more likely than Alberta to underwrite projects like a farmer-owned grain elevator company. One thing is clear, though: in the eyes of those running both Alberta and Saskatchewan, government had an important role to play in developing the region and ensuring the welfare of its citizens.

When it came to social policies, both adopted many similar ones at first. Women in both provinces got the vote in 1916, while referendums prohibiting the sale of alcohol passed in 1915 in Alberta and the following year in Saskatchewan. Note that while Prohibition eventually was discredited in Canada as well as the United States, at the turn of the twentieth century it was seen by many as a reform that would protect women and children, and civilize a sometimes rough society.

Wallace Stegner, whose description of the Prairies is quoted above, arrived with his family in Saskatchewan during this period. He tells of riding from the Canadian Pacific main line with his mother and little brother in a stage coach driven by a cowpuncher from Montana to the land they were homesteading in the Cypress Hills. "I rode the sixty miles on Buck Murphy's lap, half anesthetized by his whiskey breath, and during the ride I confounded both my mother and Murphy by fishing from under his coat a six shooter half a big as I was."[23]

Murphy, Stegner goes on, was shot and killed not long after by a Mountie, and the two incidents for a while made him think that he'd grown up on a gun-toting frontier. But it wasn't. Murphy had crossed the border for reasons that might have interested a Montana sheriff, but he carried his pistol inside his coat because Canadian law forbade the carrying of side arms. "In the American west men came before the law, but in Saskatchewan the law was there before the settlers, before even cattlemen, and not merely law but law enforcement. It was not characteristic that Buck Murphy should die in a gunfight, but if he had to die by violence it was entirely characteristic that he should be shot by a policeman."

Stegner's comment is something we'll return to in the next chapter, about the United States and Canada. But all was not rosy north of the border. Farmers in the Canadian West considered themselves very hard done by because they had to deal with the power of railroads, grain companies, and banks, all of which were based in Central Canada. They protested by sending sixty-four members of the Progressive Party to the House of Commons in Ottawa in the 1921 election in hopes that members of parliament (MPS) sworn to protect farmers' interests would make a difference. On the provincial level, though, Pitsula says that election year was the point when Alberta and Saskatchewan began to part company.

— — —

Both provinces were profoundly affected by populist movements. In recent years, populism has gotten a bad rap because many of the politicians espousing it have been megalomaniac strongmen. Oh, they may have started out with sincere desires to help the common folk—which is what populism is supposed to be about—but they succumbed to the temptation of believing that what was good for them was also good for everyone. "Populist politics," writes André Munro in the *Encyclopaedia Britannica*, "revolves around a charismatic leader who appeals to and claims to embody the will of the people in order to consolidate his own power."[24] In the current North American context, there is no better example than Donald Trump, the forty-fifth president of the United States.

In Alberta, one unabashed populist voice was that of Henry Wise Wood, president of the United Farmers of Alberta from 1916 to 1931, who advocated "group government." The idea, which sounds very much like the corporatism proposed by Benito Mussolini in Italy, would have farmers, organized labour, and professionals enter politics as economic classes. This is exactly what big business had done, Pitsula says, "namely organize to protect their class interest." The UFA formed the government in Alberta in 1921 and held office until 1935, becoming the first of a string of "one party democratic" governments that ruled Alberta until the twenty-first century. The next was Social Credit, based on the economic theories of C.H. Douglas (no relation to Tommy) that proposed giving an income supplement to all in order to boost consumption and therefore demand. Its great proponent was William ("Bible Bill") Aberhart, whose radio sermons reached across the Prairies to as many as 300,000 people. Along with the idea of underwriting consumption by ordinary folk—the exact opposite of the trickle-down economics espoused by the Right today, by the way—the Social Credit Party and Aberhart sided wholeheartedly with religious fundamentalists. Bible Bill was skeptical about any attempt "to build the Kingdom of God on earth." Certainly he would never end a speech by calling on his listeners to "build a new Jerusalem in this green and pleasant land," as Tommy Douglas frequently did.

In Saskatchewan, however, the same agrarian dissatisfaction took a turn to the left, which is where Douglas came in. A Baptist minister, he arrived in Weyburn, Saskatchewan in 1930, just as the effects of the Great Depression combined with local dust bowl disasters. Pitsula says Douglas "immediately became involved in assisting the unemployed and trying to improve the relief system." Three years later he joined the newly formed CCF, attracted by its plan to eradicate capitalism and put in place "a full program of socialized planning which will lead to the establishment in Canada of the Co-operative Commonwealth." From 1932 to 1944 a progressive farmer-labour coalition was a force in Saskatchewan politics, with the CCF winning a majority in 1944. The party also elected MPs to the House of Commons and did well in British Columbia in 1941 and 1953, as well as in Ontario in 1943.[25]

The success of the CCF attracted attention far and wide. As American sociologist Seymour Martin Lipset wrote in his landmark work *Agrarian Socialism*, nowhere else in North America had a socialist party done so well. Canada's powerbrokers viewed the CCF's success as a bigger threat than Social Credit's, and this resulted in "unrelenting red-baiting," according to John C. Conway and Aidan Conway.[26] It subsequently moderated its platform to become "a moderate social democratic party and the 'natural' governing party of the province." They note that between 1944 and 2012, the CCF/NDP won 12 of 18 provincial elections, governing for 47 of 68 years.

Lipset, whose book was published some ten years after the CCF came to power in Saskatchewan, was impressed with the way it introduced health insurance in 1947. But rather than lauding the move as the next logical step in a social-democratic program, he said the move was a response to a growing lack of support during a time when revenues were more than adequate and many of the reforms the CCF had brought in had become commonplace. "To some degree, this innovation can also be related to the CCF's gradual loss of strength. Party leaders felt the need to find an issue around which they could rally their increasingly apathetic supporters, one that had also been historically popular among the farm population," traditionally the CCF's strongest constituency.[27] Lipset added that years in power, with the responsibility of managing the several projects brought about by CCF reform—grain elevators and food cooperatives, for example—also brought a change in attitude. A CCF administrator had begun to view many of the projects as a businessman did, wrote Lipset: "He too must show a profit to his stockholders [the tax-conscious farmers], avoid wage increases if possible and raise prices."[28]

But, Conway and Conway argue, the CCF/NDP "positively transformed the lives of citizens in ways never before believed possible"—and not only in Saskatchewan, but elsewhere in Canada. Threatened by the popularity of these progressive policies, "the established pro-capitalist parties adopted features of the CCF's welfare state program," they note.

In the 1970s, the NDP government in Saskatchewan took on industrial interests by bringing half of the potash industry into public ownership, setting up a Crown corporation in the oil industry, and, as Conway and Conway put it, "impos[ing] public ownership shares on future developments in natural resources."[29]

The momentum was broken in 1982 when the NDP was swept out of power by the provincial Progressive Conservatives under Grant Devine. Times were good but discontent with Ottawa was great, and the NDP, which had supported the Liberal government of Pierre Trudeau on many issues, took a bashing, partly because, unlike the federal Liberals, they were a target close at hand. Cuts in social and health programs, and then privatization of public Crown corporations, followed. According to Conway and Conway, it was at this point that Saskatchewan's important role as a model of progressive political change ended.

During his two terms as premier, Devine outspent previous NDP govern-ments and gave tax breaks to industry, bringing the province close to bank-ruptcy. When the NDP won power again in 1992, it also took up neo-liberal policies, selling off Crown corporations and no longer supporting the social programs previous CCF/NDP governments had so proudly and effectively put in place. Conway and Conway are scathing in their criticism of NDP premier Roy Romanow, saying that his cuts went "farther [and] faster than Devine ever dared," all for the benefit of business interests.[30] Romanow's supporters, however, say that he had no choice if the province was to avoid financial ruin. According to David McGrane, "a combination of the Devine government's low taxation policies and slow economic growth...led to decreased revenues for the provincial government in the 1980s. Despite such depressed revenues, the Conservatives continued to overspend...leading to a substantial deficit." The Romanow government reacted when the NDP was back in power by moving quickly to reduce the deficit "through increases to the province's corporate income taxes, higher consumption taxes (sales, gas and tobacco), a surtax for high income earners and spending cuts."[31]

There's a huge irony here of course, because it wasn't a left-wing party that did such damage to the province's finances—Tommy Douglas was proud of not running deficits—but a right-wing one.

McGrane compares the Romanow-era government to "third way social democracy" as practised by the British Labour Party under Tony Blair in the mid-1990s, and writes of a "right wing" that had emerged in the Saskatchewan NDP beginning long before Romanow took office. Romanow's successor, Lorne Calvert, piloted the party to a narrow victory in 2003 (30 seats to the centre-right Saskatchewan Party's 28), but the NDP tanked in the next election when the Saskatchewan Party under Brad Wall won decisively. After that, Wall won twice more, but what has transpired, critics like Conway and Conway say, is not that much different from what the NDP had been doing in its last years in power, when it assured the world that Saskatchewan was "open for business." Wall stepped down in 2018, but this pro-business commitment continues. His successor, Scott Moe, who will face electors in 2020, is almost jubilant in his opposition to the tax on carbon introduced in 2019 by the fed-eral Liberals.[32] Certainly, Tommy Douglas might shake his head over what has been happening in the province with which he is so closely linked.

Conversely—and maybe perversely—in Alberta there has been a movement away from the principles espoused for eighty years by small-c conservative governments under various party banners. In 1953, at the same time that Lipset was writing about Saskatchewan, C.B. Macpherson argued in his book *Democracy in Alberta* that voters in that province have always preferred a more or less stable non-partisan business government to a climate in which two or more parties fight it out.[33] In such a regime, conflicts take place within the party—whatever it is called—not on the hustings.

But something surprising happened in 2015 in Alberta: the provincial NDP won a majority, beating out the Progressive Conservatives and the similarly right-wing Wildrose Party. That the Right was divided was undoubtedly a factor. So was the fact that after forty-four years of Progressive Conservative government, unrest rocked the party: PC premier Alison Redford was forced to resign a year before the scheduled provincial election over expense scandals. The sudden, nearly catastrophic drop in the price of oil that year didn't help the PCs either. In a province whose economy is fueled by the oil patch, the increase in unemployment was dramatic, and a lot of people were ready to vote for a party representing a different path.

Danielle Smith, a former leader of the Wildrose Party who crossed the floor to join the Progressive Conservatives shortly before the election, professed no surprise when the NDP won the 2015 provincial election. In a piece in the *Globe and Mail* she argued that many of the policies championed by the NDP in the past were fully supported by Alberta's Progressive Conservative governments.[34] Her subtext was, apparently, that Wildrose wouldn't have done this. The suspect programs included generous funding of Medicare—Alberta charged no per capita premium although several other provinces do—as well as substantial support for public education at all levels. The province has no sales tax, but royalty revenues from the oil patch until two years before the election had provided more than enough money to support these programs while balancing the books. Therefore, Smith suggested, voting NDP did not represent a leap into the unknown for voters because they were already accustomed to government playing an important role in their lives.

Writing in *Maclean's*, Colby Cosh had a similar take on what happened. His analysis also brings to mind both the effect of people moving because they are unhappy with a place and the work of demographer Robert Cincotta, who predicted that Tunisia would see a better outcome from the Arab Spring than other countries because of the role that a population's median age plays in its politics:

Past Conservative election efforts have depended heavily on the fact that hundreds of thousands of older Albertans are refugees, or the children of refugees, from NDP-run provinces, particularly neighbouring Saskatchewan....

But those folk memories are fading. Alberta's median age is near its expected peak, and the province is thus experiencing the transition from Baby Boomer dominance to the demographic bulge in under-40s.... [For them], the New Democrats are just socially and environmentally concerned people who deserve a fair shake. They have already played an enormous role in the sweeping electoral victories of Calgary Mayor Naheed Nenshi and Edmonton's Don Iveson.[35]

In the 2019 provincial election, the stars were not aligned in the same way, however. The Right was united under a new banner, the United Conservative Party, led by Jason Kenney, one of Steven Harper's right-hand men in the federal government from 2006 to 2015. Consequently, Rachel Notley and the NDP came in second despite the province's changing demographics.

Nevertheless, the Alberta of Stephen Harper appears to be changing as much as the Saskatchewan of Tommy Douglas has. Cosh's reference to the changing population may be particularly pertinent—and a harbinger of things to come. According to the 2016 Canadian census, immigrants made up 10.2 percent of Saskatchewan's population and 21.2 of Alberta's. This represents a jump from five years before, when the figures were 6.8 percent for Saskatchewan and 18.1 for Alberta.[36] Like yeast in bread dough, immigrants can transform things, including the political culture—something we'll discuss more thoroughly in the next chapter.

So what's ahead for Alberta and Saskatchewan? Hard to tell. In the meantime, the wind still blows through the grass, the grain still grows tall, and the geographic boundary between the two provinces remains difficult to discern.

CANADA AND **THE UNITED STATES OF AMERICA**

When I think back to the Sunday in 1968 when we crossed the US-Canada border, I try to remember what felt different about our new country. What we saw in Detroit probably sums it up. A little more than a year earlier a large part of the city had burned during a five-day race riot that started when police raided an after-hours drinking spot. By the end of the conflict—sometimes called the Detroit Rebellion—43 persons were dead, 1,189 injured, and more than 7,000 arrested. The material damage was also substantial, amounting to between $287 million and $323 million in 2016 dollars.[1]

We knew this, of course—although it was only when I started working on this book that I came across home movies of the troubles that look startlingly like those shot by US soldiers in Vietnam (and which you'll recall from the chapter on the two Vietnams).[2] On the Sunday morning that we arrived in Detroit, we got off the freeway with the idea that we'd find Wayne State University, where a friend of my husband's from grad school was teaching. Not that we intended to drop in and say hello: the sun was just coming up after all; we just wanted to see what the neighbourhood looked like. But we quickly lost our way and found ourselves driving down deserted streets lined with burned-out buildings and piles of rubble. In the end we must have stopped at a store or service station to ask directions—it's something of a blur all these years later. Eventually we made it downtown, where, still a little shaken by what we'd seen, we went into the nicest hotel we could find to eat breakfast and wash up a little (we'd driven all night, you may remember, because there was no room in the inns around Chicago).

The breakfast was good and the welcome cordial. It helped that a professional baseball team was in town to play the Detroit Tigers; my husband with his crew cut looked like he might be a baseball player. An older man at the next table even told him that "his" team had played well the day before. Then when we left we found a black man with a shopping cart and a sleeping bag over his shoulders crouching just outside the hotel's grounds: homeless, probably, and something I hadn't encountered before. He got up quickly and took off when we approached.

The Ambassador Bridge between Detroit and Windsor is the most travelled border crossing in the North America—and an important link between two very different countries. *Source: Flibirigit, public domain.*

The contrast couldn't have been greater. Canada, we were sure, didn't have problems like the ones that produced the Detroit riots. Windsor, just across the Detroit River in Ontario, was cleaner, we saw immediately. There were no burned-out buildings. Folks were polite in the working-class neighbourhood where we stopped on the way out of town to buy provisions for lunch. "Pardon me," the young man behind the counter at the convenience store asked when he didn't catch what the person in front of us asked him. "Sorry," someone else said when he brushed passed us on the way out.

We would soon learn that slavery had been legal in Canada until the 1830s; that generations of Indigenous Canadians have suffered from policies that many qualify as genocide; that prejudice continues against African-Canadians and other people of colour; and that homelessness would become common-place in Canadian cities. Yet the viciousness of American racism was absent.[3] It's telling that the Americans had to fight a war to end slavery, but in Canada it was outlawed after debate in the British Parliament: the bill passed nearly thirty years before Abraham Lincoln decreed the Emancipation Proclamation in 1863.

What seemed most important to us at the time, however, was the way the Canadian officials we spoke to viewed Vietnam: the US draft was a matter between a man and the US government, they all said—Canada wasn't part of the equation. The country, which had sent men to fight in Korea, was not going to do that this time around.[4] Prime Minister Lester Pearson had won the Nobel Peace Prize for his role in setting up a United Nations Peacekeeping force in the Middle East following the Suez Crisis, and Canadian soldiers were an important part of that effort. Canada was not merely a clone of the United States when it came to foreign policy.[5]

We continued. We reached Montreal on the afternoon of a hot, muggy Labour Day, going down Sherbrooke Street past white-clad lawn bowlers in Westmount who seemed the antithesis of the poor man outside the hotel in Detroit. There were no road signs in kilometres, no temperatures in Celsius, few signs in French—that would come later as Canada changed—but it was clear that we sure weren't in Kansas anymore, Toto.

— — —

The border between the United States and Canada has never been as conten-tious as the one between the United States and Mexico became after Donald Trump was elected president in 2016. Rather, it has frequently been called the longest undefended border in the world. On the north side lies a country with much colder average temperatures, a smidgeon more area, and something more than a tenth of the population: Canada measures 9,984,670 square kilometres compared to 9,826,630 square kilometres for the States, while in January 2019 Canada had a population of about 37 million compared to the 328 million in the States.[6] Supposedly Americans are more individualistic, violent, litigious, innovative, and open, while Canadians are more peaceful, polite, concerned about the collective welfare, and more inclined to follow orders. The received wisdom would have it that the two countries' mottos say it all: "life, liberty and the pursuit of happiness" versus "peace, order and good government."

But the differences between the two countries go much deeper. At this writing, life expectancy, infant survival rates, income equity, social mobility, and school children's rankings on international educational tests are all higher in Canada than in the United States.[7] On the other hand, if you want to own a weapon, pay lower taxes, or pass your winter in a place usually less snowy, the United States is the place to be.

Until recently, probably not many Americans wondered about the reasons behind those differences. Ignorance about Canada is high despite increased interest in moving north among people upset that Donald Trump is president.[8] Canada is more likely to be part of the punchline for a joke: as comedian Robin Williams put it, "You are the kindest country in the world. You are like a really nice apartment over a meth lab."[9]

Or perhaps Canada is seen as a place to escape to in times of trouble, and not just of the recent, Trumpian kind. Two stars of twentieth-century American literature—Philip Roth in *The Plot against America* (2004) and Sinclair Lewis in *It Can't Happen Here* (1935)—portrayed Canada as a place where people could get away from the clutches of a fascist/brutal/racist government that has taken over the United States. Canadian Margaret Atwood did the same thing in her novel *The Handmaid's Tale* (1985), but perhaps that attitude should be expected from a Canadian nationalist. It's worth noting that the three novels were written at three different points in time over seven decades, which suggests that malaise about American politics has a long history.

Why that is, why Canada isn't part of the United States, and why it has gone its own independent way, are questions of immense importance for Canadians—and perhaps to the world. Relations between the two countries are one-sided though. As Justin Trudeau's father Pierre said of the United States when he was prime minister, "Living next to you is in some ways like sleeping with an elephant. No matter how friendly and even-tempered is the beast, if I can call it that, one is affected by every twitch and grunt."[10]

In my experience, Americans frequently assume that Canadians would jump at the chance to become part of their neighbour to the south. My own parents never quite understood why we left the United States or, worse, why we stayed in Canada, particularly since a generation before segments of the family had travelled the other direction. But many Canadians bristle at the idea that they'd be better off as an appendage of the United States: indeed, an early reader of this book suggested that there'd be uprisings across the country if the United States flexed its muscles and tried to move north.

— — —

Yet invade is what Americans have done several times—and that's not counting a few disputes that arose during the process of determining just where that long border runs. The first attempt goes back to before the actual founding of the United States, when what is now Canada was just another part of British North America. Early in the American Revolution, the upstart Yankees headed northward to fight British forces in what was then the Province of Quebec, which had been taken from the French only a decade before. Not only were Quebec City and Montreal strategically important, but the revolutionaries also believed that the French-speaking population would rise up against their British conquerors and join the rebels.

You'll remember that when Ethan Allen and his Green Mountain Boys took Fort Ticonderoga on Lake Champlain in May 1775, the victory was one of the first in the Revolutionary War. Then, during the summer, Allen and his men continued north, reaching the south shore of the Saint Lawrence, across from Montreal, by September 24. They crossed the river in three boats and dug in for a battle. When British forces arrived the handful of Quebecers Allen had managed to recruit to the revolutionary cause took off, and when the dust settled, Allen was captured. More-disciplined American forces took Montreal in November, however, and for the next six months the rebels governed the city from the Chateau Ramezay, the governor general's residence.[11] But the American occupation ended after another Yankee invasionary force, this one coming up the Chaudière River from Maine, was defeated when it tried to take the British stronghold at Quebec City on December 31, 1775.

American leaders were surprised by the lack of support from the locals: they assumed that the French-speaking population would greet the invaders with open arms and join them in a grand effort to push the British off the continent.[12] But they made no secret that many Americans were upset about provisions in the Quebec Act, passed by the British Parliament in 1774, giving Roman Catholics the right to worship as they wished and to hold political office. Some Yankees were afraid this would open the door to more Catholicism, something that would be anathema to many in the new, largely Protestant nation. Not surprisingly, when news of this attitude got out, it did not elicit enthusiasm from *les Québécois*. It was the kind of American misreading of Canadian mores and ideas that has happened several times since. It also points up the role that language often plays in determining the fates of nations.

The second conflict between the United States and what would become Canada was the one that Donald Trump invoked in 2018 when he slapped tariffs on Canadian steel and aluminium imports for "security reasons." "Didn't you guys burn down the White House" in the War of 1812 , he reportedly snapped at Canadian prime minister Justin Trudeau when the two talked on the telephone about Trump's salvo in the new trade war.[13] The truth is that Canada as such didn't exist then. Fighting in the two provinces of Upper and Lower Canada (more or less Ontario and Quebec, respectively) was part of a larger conflict between Great Britain and the fledgling United States, and the attack on Washington was mainly an endeavour of the British navy. Nevertheless local militia in the two Canadian provinces played a prominent role; and once again the Americans mistakenly thought they'd be welcomed as liberators in the Canadas—former president Thomas Jefferson, for one, was quoted as saying that taking them was "a mere matter of marching."[14] The end result was the restoration of the boundary between the United States and British North America fixed at the end of the Revolutionary War in 1783. Not quite a Canadian victory, but the war did keep the Americans from claiming Canadian territory for a while.

It should be noted that Canada was twice invaded by forces from the United States, whose goals went beyond simple conquest of territory. In 1838 a group of Canadian patriots invaded from New York State in the final stages of the nearest thing Canada has had to a revolution, the Rebellions of 1837–8. The commander of the invasion force, Montreal physician and politician Robert Nelson, declared the Republic of Lower Canada before being beaten by British regulars.[15]

Later on in the century, Irish nationalists in the United States, the Fenians, attempted to take Canada by invading Quebec and Manitoba with the goal of trading the country for an independent Ireland free of British rule. They were decisively turned back, partly because the Irish in Canada, who the Fenians thought would flock to their cause, tended to be from the Ulster counties, and weren't keen on an independent Ireland.

Then in the fallout from the Civil War, Americans wanted to use Canada as a chip in negotiations with Great Britain over reparations for damage done to us warships by the British navy. American negotiators wanted an incredible $2.125 billion, and added that they'd take Canada in exchange if they couldn't get the money. In the end they settled for much less, but the agreement, the Treaty of Washington, marked a step forward in Canadian affirmation since it was signed in 1872 by the us and British governments, as well as—a first—by the fledgling Canadian government.[16]

(Whether the United States would have known what to do with Canada if it acquired it at that point is a huge question. In the 1871 census, 30.1 percent of the Canadian population said they were French, and these million or so Francophones would have been hard to assimilate into the United States, whose working—but not official—language has always been English. A hint of what might have happened can be seen in relations between Puerto Rico and the United States. The largely Spanish-speaking territory, acquired by the Americans in 1898, has never been made a state, even though Puerto Ricans became American citizens in 1917,[17] and when Hurricane Maria struck a hundred years later, many charged that they were treated as second-class citizens.[18])

Canada's foreign policy remained that of Great Britain until 1931, when the Statute of Westminster allowed it to enter into agreements with other countries on its own. The close relationship with Great Britain continued, though, and concern about that connection underlies an audacious plan formulated by the United States in the 1930s to invade Canada. War Plan Red presumed that Canada, should the United Kingdom be defeated in a war, would feel great pressure to follow orders from forces hostile to the United States. To counter that, the plan called for a naval takeover of Halifax, followed by an armoured column striking north from New York and Vermont to take Montreal and Quebec City, just as Yankee forces had done during the Revolutionary War. Columns would also move on Toronto, Niagara Falls, Winnipeg, and Vancouver.[19] Reading accounts of the thinking that went into the plan, you can only shake your head. Once again the Americans didn't understand what was going on north of the border: when war in Europe did arrive, Canada entered the fray against the Axis Powers with its own declaration of war two years before the Americans did.

— — —

What exactly is the root of the differences between Canada and the United States? It's clear it has little to do with the border itself, which is largely a reflection of arbitrary choices. The western part—2,092.155 kilometres (1,300 miles) along the 49th parallel—was agreed between the young United States and Britain in 1818, sight unseen. Nobody had the foggiest idea of what kind of terrain it went through, and it was decades before it was surveyed. When it was, the surveyors' marks were frequently moved, or—since many of them were wooden—burned for fires on the range. But while the exact route might be hard to discern, it loomed large in the history of the West. The Sioux, chased from their hunting lands in the United States, called it the Medicine Line because the US Cavalry had to stop—even in hot pursuit—since they weren't authorized to pursue Indigenous people across the border.[20]

The story in the East was similar, with the final boundary line not determined until the late nineteenth century. A long stretch was arbitrarily fixed at the 45th parallel separating Quebec from New York and Vermont. That it passes through one small town—Derby Line, Vermont, a.k.a. Stanstead, Quebec—bears witness to the fact that the final boundary was determined well after settlement of the region began in the late 1700s.[21] Where Quebec meets New Brunswick, the border follows the height of the land, then drops due south to the St. Croix River, a boundary that was only agreed on after some sabre-rattling in the 1840s that was called, with marvelous colonial exaggeration, the Aroostock War.[22]

Today English is the mother tongue for most citizens of both countries according to the most recent censuses: 56 percent of the population of Canada in 2016 and 80 percent of the population of the United States in 2011.[23] But a major difference between the two countries lies in their second most popular languages and their histories. In Canada, French was the mother tongue of 21 percent while in the United States Spanish came second, at 10.6 percent.[24] The idea of Two Founding Peoples—the French and the English—is at the heart of Canadian history, but the large number of Spanish speakers in the United States is something relatively recent, even though for a couple of centuries Spain and later Mexico controlled a quarter of the present-day United States.

About 100,000 Mexicans lived in the territory between Texas and California that Mexico ceded to the United States after it lost the Mexican War in 1848. They had the option of becoming American citizens, but not many did. In any case their numbers were miniscule—only 0.4 percent of the 23 million persons that the US Census recorded in 1850. By the time the twentieth century rolled around, Spanish speakers accounted for only about 0.15 percent.[25]

Times have changed: for reasons we'll go into a bit later, about 11 percent of US citizens currently are of Mexican descent, and another 5.7 percent are Hispanic with origins elsewhere in Latin America.[26] Nevertheless, despite their growing numbers, Hispanics do not occupy anywhere near as large a place in the US political landscape as French Canadians do in Canada, nor have they been as influential.

— — —

French and/or Quebec institutions are the root cause for major differences between the United States and Canada today, according to American sociologist Jason Kaufman. He argues that these differences go way, way back, to the strategies the British and the French used to colonize North America.[27] The British had three models for their colonies: one that set them up as private, for-profit corporations (for example, the Province of Massachusetts Bay, which included Maine, Connecticut, and Rhode Island); a second in which the colonies were private estates granted to one or a few proprietors to do with as they wished, in a pattern much like that of the Northern Ireland plantations (as in New York, New Jersey, Pennsylvania, Georgia, the Carolinas, and Maryland); and a third containing "royal colonies" where the king retained direct ownership of the land (New Hampshire and Virginia).[28]

The three types of colonies differed in the way they allotted land to settlers, but in all cases the presumption was that Europeans could claim the land without compensating or even consulting the Indigenous population. Before the end of the Seven Years' War in 1763, Britain had no colonies in what would become Canada, but from the early 1500s France had been settling the Saint Lawrence Valley. Its colonial policy was different from that of the British. In essence, the French Crown retained control of all the land—in a system that Kaufman calls "flexible feudalism"—with an orderly distribution of farmland along the water courses. Indigenous people weren't consulted here either, and there were many armed skirmishes. Seigneuries—land grants given out to quasi-noble individuals—were divided into long, thin plots, fronting on a river or, later, as the land began to fill up, along a roadway.

Flying into Montreal today you can see the pattern from the air, in marked contrast to the rectangular farms to the west in Ontario and beyond. Montreal's streets themselves echo the practice, as the first grids were set up in reference to the Saint Lawrence. The result is long, thin blocks aligned not on compass headings but several degrees off. What locals call north is really north-of-west, which means the sun on June 21 appears to set at the end of "north-south" streets. That was just one of the things we had to get used to when we first arrived in Montreal.

Our reason to come here was originally economic—a job for my husband at McGill University. Economics was the reason that New France was established in the first place too. The French hoped to find gold and silver, as the Spanish had in South America, but that never happened under their watch. "Diamonds" brought back by the first French explorer, Jacques Cartier, turned out to be quartz crystals, and "faux comme un diamant du Canada" became a common insult.[29] But from the beginning, the French were able to capitalize on another major resource, furs, particularly beaver pelts. A sizeable proportion of New France's male population spent months every year in the woods, either maintaining their own trap lines or trading with Indigenous people for furs.

Early French settlement and trading took place against a background of war among various Indigenous tribes. The Five Nations of the Iroquois were intent on dominating the region that became New France and New England, much as the various European powers were themselves fighting for control of that continent. Explorer Samuel de Champlain earned the enmity of the Iroquois in 1609 on his first trip up the Richelieu River and south on Lake Champlain when two Iroquois warriors were killed by the French and their escort of sixty Algonquin.[30] Iroquois raids on French settlements in the Saint Lawrence Valley followed over the next thirty years. But by the end of the century it became clear that coexistence among the various Indigenous nations and the French would be mutually advantageous. The Great Peace of Montreal was negotiated and signed in 1701 by forty representatives of various tribes and the French establishment. This led to relatively cordial relations between the Indigenous population and the French until the end of the Seven Years' War.[31]

You'll remember from the chapter on Vermont and New Hampshire that in 1704—only three years after the Great Peace—a joint French and Indigenous force raided the village outpost of Deerfield in the Massachusetts colony. It was just one of many such incidents on the New England frontier that was matched by conflict between English settlers moving into Indigenous lands farther south in the Appalachians. Kaufman says this continuing state of armed uncertainty left a lasting mark on the American psyche. Mark Cronlund Anderson goes further in his Holy War: Cowboys, Indians, and 9/11s. He asserts that because of the American frontier experience, the country's founding myth insists on violence to overcome the forces of darkness, which may be defined as the Indigenous population, other nations, or rogue individuals.

Furthermore, the very existence of these "enemies" is construed as an aggression and so pre-emptory attack is justified. Invoking the concept of holy war in cases like these is not hyperbole, he says.[32]

The large number of immigrants to North America from Scotland and Northern Ireland—the Scotch-Irish, as they are called in the United States—would appear to have reinforced this attitude. Historian Arthur Herman, in his *How the Scots Invented the Modern World*, says that the first wave of immigration from Scotland, begun in 1717 after a failed harvest, was of feisty Lowland Scots and Ulstermen. In the southern Appalachians "the habits of colonizing Ireland and seizing arable land from Catholic enemies carried over to the New World."[33] The Scots' "insatiable desire for lands, and the willingness to fight and hide to keep it, laid the foundation of the frontier mentality of the American west."[34]

The term "rednecks," Herman says, was a Scots border name for Presbyterians, while "cracker" comes from the Scots word *craik* for talk, and came to mean a loud talker or braggart. He adds that Ulster Scots had a long-standing hatred of the English and so were natural rebels during the American Revolution: "the most God-provoking democrats this side of Hell." Perhaps half of George Washington's army at Valley Forge were Ulster Scots, he says. Many of descendants of these rebellious Scots can be found among the militant right-wingers of today's Trumpism, including the hard core of white, non-college-educated men who pride themselves on being bedrock Trump supporters.[35]

In contrast, more-educated Scots from Edinburgh and Glasgow were active in trade in Britain's North American colonies, while Highland regiments had also been stationed there from their beginning. When the revolution was over and the thirteen colonies had formed the United States, many of these people went north to escape what they perceived to be a dangerous new government. Many were rewarded for their loyalty to the Crown with land grants in Upper and Lower Canada. Both Seymour Martin Lipset (he of the landmark study of Saskatchewan discussed in the previous chapter) and, later, Kaufman say that this movement was extremely important in setting the tone for Canada—and for the United States as well. As in Carthage, you got rid of those who don't agree with you by encouraging them to move on.

A direct result of the violent relations between settlers and Indigenous people on the frontier is that persons who claim Indigenous heritage today make up a considerably smaller proportion of the US population than they do in Canada. In 2011 the figure was 4.3 percent in Canada, while in the United States a decade before it was just 1.5 percent.[36] In effect, the Americans were more "successful" in emptying the country of its Indigenous population. Whole groups were effectively exterminated, while others, particularly in New England and the West, were chased north across the Medicine Line.

Canada's treatment of Indigenous people has been, and still is, shameful too. Today, suicide and disease rates in First Nations communities are much higher than elsewhere, and charges of systemic racism against the Indigenous population are growing louder.[37] The Canadian federal government is responsible under the Indian Act for education, health care, and much more, but far too frequently it has failed to keep its promises. The role of residential schools, to which Indigenous children were sent for more than a hundred years, has come under intense scrutiny. The intent was to assimilate as well as educate Indigenous children, but "cultural genocide" is what resulted. Horrific stories of sexual and physical abuse perpetrated by those who ran the schools have become commonplace.[38]

South of the border the Indigenous population's current condition receives some attention, including investigation into the residential schools established there even before the ones in Canada.[39] But the United States has another race problem that takes up an enormous amount of space in its history: that of the enslavement of people imported from Africa. Slavery was the economic foundation of many of the thirteen colonies, it was the professed reason for the Civil War, and while Americans elected an African-American president twice, tension between black and white in the States remains high.

Across the Medicine Line, people of colour say Canadian society is also riddled by racism, but a measure of how the two countries differ when it comes to the depth of prejudice and racial violence is reflected in the fact that between the 1880s and 1968 one person—significantly, an Indigenous man— was lynched in Canada, while in the United States 3,446 black people were lynched. (So were 1,297 white people, but it appears that most of them were lynched for helping blacks or for being anti-lynching.[40])

Even though Canada was a safe haven for slaves in the years before the American Civil War because slavery was outlawed in British North America in 1834, the number who followed the Underground Railroad north totalled only between thirty and forty thousand.[41] Canada's African-descended population grew slowly until the mid-twentieth century, when immigration from the Caribbean and Africa increased greatly with the introduction of a new point-based immigration system. Now, the majority of Canada's black population has relatively recent roots in the Caribbean and Africa, rather than in earlier migrations.[42] But still the black population is proportionately much larger in the United States than in Canada, 13.4 percent compared to 2.9 percent in Canada.[43]

— — —

While slavery left deep scars in the United States, Kaufman, like Cronlund Anderson, insists that it was constant and early confrontation with the Indigenous population on the frontier that established a pattern of violence in American society. Militias to fight Indigenous people were among the first community organizations formed by Americans, he points out. Therefore it is no coincidence that the Second Amendment to the US Constitution adopted in 1791 reads: "A well-regulated militia being necessary to the security of a free state, the right of the people to keep and bear arms shall not be infringed." This carry-over from the frontier mentality is used as a justification by a sizeable and influential part of American society to refuse to do anything about regulating firearms today. Even though Canadians aided the United States in setting up the National Rifle Association and a militia system shortly after Confederation, and they still own a lot of guns, attitudes toward firearms are considerably different in Canada.[44]

Indeed, one of the biggest differences between the United States and Canada today is closely linked to Americans' passion for firearms: the per capita homicide rate south of the border has been about three times that of Canada each year for the last fifty years.[45] Gun culture in general has far more political clout in the United States than it does in Canada, with the American NRA successfully blocking nearly all attempts to regulate firearms in the United States. Even the string of mass shootings—some by simple crazies, some by politically motivated terrorists, some by out-and-out racists—that has marked the last few decades has not prompted enough reaction to get gun regulations passed by the US Congress. In contrast, the massacre of fourteen young women in 1989 at the École polytechnique in Montreal brought about a mass movement for a federal long-gun registry. It existed for more than twenty years until it was abolished by the government of Conservative prime minister Stephen Harper against the advice of the Canadian Association of Chiefs of Police.[46] Not surprisingly to anyone who is familiar with the relative peacefulness of the pattern of settlement in the Saint Lawrence Valley, the province of Quebec set up its own long-gun registry in January 2018. While there currently are few regulations regarding handguns and assault rifles in Canada, the Trudeau government has put forward a bill that would outlaw them both: widespread consultations were scheduled before the October 2019 federal election.

But it should be understood that being prudent about firearms doesn't mean military history is unimportant in Canada. It is, it's just that the lessons drawn are different. One of our greatest shocks when we arrived was seeing Canadians, young and old, buying and proudly wearing red poppies in the run-up to November 11, called Veterans' Day in the United States and Remembrance Day in Canada. Back in the States at that point, no one who questioned the wisdom of the American war in Vietnam would ever wear one. Sold by the American Legion, to people like us the poppy was a symbol of blind patriotism.

But in Canada it is nothing of the sort. The poppies as a symbol have their origin in the poem "In Flanders Fields," written in May 1915 by John McCrae, a Montreal doctor and artillery officer.[47] The United States would not be in that war for another two years, but Canadians had been on the ground since shortly after the first shots were fired. For the first time, they played a big role on the world stage, and the experience is remembered by most Canadians with considerable pride. That pride grew in the next world war, when Prime Minister William Lyon Mackenzie King recalled Parliament to declare war on Germany a week after Britain did. The aim of the delay was to declare Canada's distinctness and independence: an important step toward international recognition as a nation.[48]

Note that Canadians did not have a revolution in the American sense. The United States promulgated a unilateral Declaration of Independence, and created a constitution elaborated by a group of men more or less elected by their peers. The nation itself was forged in battle, as Abraham Lincoln put it, during the Revolutionary War. Canada, in contrast, was formed ninety years later when British colonies came together in an orderly and amiable fashion. Indeed, Britain seemed almost happy to be done with governing those not-too-wealthy colonies across the sea. The British could henceforth devote their attention to consolidating their holdings in India, and get ready for the scramble to claim great regions of Africa.

Consider, too, that when Britain was thusly involved in extending its empire, the United States was attempting to play the same game in the Western Hemisphere. The Monroe Doctrine, formulated by President James Monroe in 1823, proclaimed that the European powers should keep out, and promised that those countries who wanted to meddle would feel the wrath of the United States.[49] Its consequences have shown up indirectly a few times in this book already—the attempt to stop piracy in the Mediterranean, for example; the Mexican War, which brought Texas, California, and much of the territory that would become Nevada, New Mexico, Utah, and Arizona, as spoils; and American interventions in the Dominican Republic and Haiti. While the territorial ambitions of the Hudson's Bay Company can be seen as a sort of imperialism, and Canadian troops fought for the British cause in the unabashedly imperialist Boer War of 1899–1902, the Monroe Doctrine had absolutely no equivalent in the Canadas, even after Confederation in 1867.

Then, in the mid-twentieth century, when the world was divided into two spheres of influence, that of the Soviet Union and the United States, the US-Canada border became less important than the one that ran along the top of the North American continent. With the advent of long-range intercontinental ballistic missiles, had the Cold War turned hot, Canada would have been the flyway for the great nuclear-bomb-carrying birds headed for the cities of either great power.

Whether to let Canadian soil be home to nuclear arms elicited great controversy: the decision to join NORAD (the North American Air Defense Command) brought down the government of Progressive Conservative prime minister John Diefenbaker. Public opinion, as well as that of the country's leaders, was deeply divided.[50] The Liberals under Lester Pearson won election as a minority government in 1963, and while missiles might possibly have been acquired then, Pearson veered away from that path. His government also prohibited nuclear arms in Canada as well as all uranium exports that were not destined for peaceful purposes. Canada subsequently became the first country with nuclear capability to renounce nuclear arms, signing the UN Treaty on the Non-Proliferation of Nuclear Weapons in 1968.

This action was in keeping with another aspect of Pearson's politics, one that has gone a long way to creating the image of the country as a "peaceable kingdom"—the establishment of a UN Peacekeeping force.

More than 100,000 Canadian troops have served with UN forces over the years. And yet, though the idea that Canadians are a peaceful people pleases many Canadians, in recent years the number of Canadian peacekeepers plummeted: in 2017 it reached a thirty-five-year low of 68.[51] Since then, Justin Trudeau's Liberal government changed things a bit. Canada deployed 250 military personnel in 2018 as a UN Peacekeeping force in Mali; not coincidentally, Canada was lobbying for a seat on the UN Security Council at the time. The new effort may not be much, perhaps, but it's a better score than before.

— — —

In many respects Justin is his father's son: when we arrived in Canada Pierre Elliot Trudeau had become prime minister only a few months before and everyone knew him as a cultured citizen of the world. He was dashing, frequently irreverent, and Liberal, which all seemed good to us (we had yet to understand the difference between Canada's Liberals and the NDP—that would come later). He also spoke elegant English and French, and Canada's French fact seemed wonderfully exotic to us.

Not that we had any appreciation then of what that meant. It was only later that we realized what profound effects Quebec's quest for affirmation in the twentieth century has had on Canada as a whole, effects that have deepened the differences between Canada and the United States.

Some observers argue that those effects are negative. Conservative thinker Brian Lee Crowley—a great admirer of Stephen Harper—holds that attempts to "appease" Quebec diverted Canada from a path that he says it had been following since before its founding. In his book *Fearful Symmetry* he argued that until the 1960s the country had an ideological commitment to family and work combined with suspicion of big government. Quebec, however, needed government intervention to further its *projet de société*, which included protecting Quebec's industry, investing in its firms, and—most importantly—regulating what language was used in the workplace and in public. The implicit threat was that Quebec might separate if it didn't get the support it needed, so the feds anted up. Crowley says the other provinces also wanted what Quebec was getting, with the result that the federal government's shadow grew much longer.

"An expanding government with myriad programs has damaged the country's work ethic and undermined the importance of family, resulting in crime, abortion, suicide, divorce, depression and a host of other social ills," Crowley told *Globe and Mail* columnist Jeffrey Simpson shortly after his book was published.[52] That assessment was not far from that of Stephen Harper's most enthusiastic partisans, and when in government, Harper's Conservatives did much to undo previous governments' social programs. Starving the beast, as the saying goes, was their watchword.[53]

This includes the social program that is probably Canada's most popular, and, according to some, its defining one: single-payer, universal health insurance. Canada's Medicare was born (as we saw in the previous chapter) in the late 1940s with hospital insurance schemes. Complete, nationwide coverage for all medically necessary health services was achieved in the 1970s: Yukon was the last of the provinces and territories to sign on, in April 1972.[54]

Since World War II, Canada and its provinces have also cobbled together a network of other social programs that provide a safety net far stronger than anything found south of the border. This is despite the fact that the United States arguably had better social programs and more income equity during the New Deal 1930s and into the wartime wage controls of the 1940s: the Canada Pension Plan, for example, was instituted a full thirty years after the United States began Social Security in 1935. But as Paul Krugman, who won the Nobel Prize for Economics in 2008, argues in his *The Conscience of a Liberal*, beginning in the 1970s, the United States saw a systematic effort by right-wing forces with lots of money to undo the reforms of the New Deal.[55] In Canada, attempts have also been made to dismantle the social safety net, but they have been much less successful.

One reason for this may be Canada's more stringent campaign-finance regulations, which have kept Mega Money out of Canadian elections. Ultra-conservative forces with deep pockets have less influence when they can't buy elections easily. Campaign finance is a topic too big to be explored here, but two examples are instructive. Since 2015, only individual Canadians—not corporations or unions—can contribute at most $1,500 a year per party, and any contribution over $200 must be published. In the United States, however, individuals can give $2,700 to candidates, and up to $107,000 to national party committee accounts, while independent expenditure-only political action committees (sometimes called "Super PACs") may accept unlimited contributions, including those from corporations and labour organizations.[56]

Whatever the reason, the result today is that Canada has greater social and economic mobility than the United States. As the Brookings Institution asked rhetorically on Canada Day 2014, "Has the American Dream Moved to Canada?"[57]

— — —

When we arrived in Quebec, it was a time of immense change. The post–World War II period saw a tsunami of attempted reforms around the world: independence movements in the developing world and the advent of the welfare state in the United Kingdom are only two examples. Quebec's Quiet Revolution—generally considered to have started at the beginning of the 1960s—can be seen simply as the province catching up with what was happening elsewhere.

We ran headlong into one manifestation of the bad old days when I went to sign the lease on the apartment we took right after we arrived. It was I who had done the looking, as I had done in California, and my husband, busy with his new job, trusted my judgment. But when I showed up at the office of the notary—a legal professional who deals mostly with contracts, and whose importance has no exact equivalent elsewhere in North America—he just laughed at me. "A married woman? You can't sign anything, you're a legal child. You're chattel property. Your husband will have to come in and sign."

So I called him, and he ran over after work to do what was necessary, while I fumed. Yet within a year the wave of reform sweeping Quebec had reached even to the dusty domain of family law, giving married women full rights to execute contracts. Ten years later, mores had changed so much that women when they married no longer had the option of taking their husband's names: they had to keep their birth one. A case in point is Justin Trudeau's wife, Sophie, who is Sophie Grégoire legally because she had to keep her birth name since they were married in Quebec.[58] (I should add that during a transition period many of my married women friends reverted to the names they were born with, but I chose to keep my husband's. After all, husband's name, father's name: it's all patriarchal!)

The late 1970s also saw the beginning of a publicly funded day-care program, which has evolved into a system providing quality day care at affordable prices for the lion's share of Quebec children.[59] Coupled with the extension of the employment insurance system throughout Canada to cover up to fifty weeks of maternity and parental leave, the number of women in the labour market has soared.[60]

Compare that with the situation in the States. President Trump promised to set up a maternity-leave package, but what was proposed was nothing like the Canadian plan. The Economic Security for New Parents Act, introduced in August 2018 by Republican senator Marco Rubio with enthusiastic support from the president's daughter Ivanka, would have merely allowed parents to borrow from their Social Security accounts to pay for six weeks paid maternity leave. It wasn't passed before the mid-term elections, and while it's possible that some other, more progressive measure might come up in the new Democrat-controlled Congress, nothing is certain.[61] Some states, however, have stepped into the breach, at least in part. California, for example, requires employers to allow employees up to twelve weeks of parental leave after the birth or adoption of a child, but benefits are nowhere near as generous as those in Canada.[62] Combine that with spotty health insurance coverage despite Obamacare (which Trump tried so hard to kill) and expensive child care when it exists, and it's clear that parents are better off in Canada.

But we weren't concerned about that when we arrived. Having kids was in the future. What was on our minds was fitting in. One somewhat unpleasant surprise was the discovery that the cohort of young professors my husband belonged to was not universally welcome. Between 1960 and 1975, the total number of full-time university teachers in Canada nearly quadrupled, from about 8,000 to about 31,000, as the country tried to keep pace with the expansion of higher education that had started after World War II in Great Britain and in the United States.[63]

Where to find all those teachers? Certainly Canadian universities hadn't yet turned out enough PhDs to fill the job slots. But the boom in higher education in the United States had begun earlier, which meant that a lot of young PhDs in the States were already looking for jobs, among them my husband. Yet even as we settled in, a movement was growing that worried about importing a flock of American professors. Would all those Americans subvert young Canadians' development and stunt the effort to build a solid, authentic Canadian identity?[64]

From the vantage point of more than four decades later, it seems that they didn't. True, Stephen Harper was influenced by ideas and teachers from south of the border. The University of Calgary, where he got his bachelor's and master's degrees in economics, was founded in 1966 during the period when so many professors were hired from the States. John Ibbitson, in his biography of Harper, says that the economics department there was started with a decidedly right-wing bent, and aimed to create a cadre of graduates who saw the world through a business-friendly outlook. Many of the faculty were recruited from American institutions that shared that orientation. The political science department had much the same inclination, Ibbitson writes: one Carleton University professor called it "the department of redneckology."[65]

But the University of Alberta appears to be the exception. If anything, the many left-wing, anti–Vietnam War Americans who were attracted to Canada were ready to embrace the mores of the Peaceable Kingdom, and moved Canada farther along the path indicated by Pearson and Trudeau. As draft dodger and folk singer Jesse Winchester sang, if you liked Franklin Delano (Roosevelt) you were going to like Pierre Trudeau.[66]

In Quebec, ideas that might be construed as left wing were particularly important during the Quiet Revolution. Government was viewed as essential in reinforcing culture—French-speaking culture, *bien sûr*. Laws making French the province's official language, as well as referendums on separation, were decades in the future, but by the time we arrived Québécois were well on their way to becoming *maîtres chez nous*, or masters in their own house.[67]

Several other quasi-governmental institutions were working to solidify Canadian identity. One was the CRTC, the Canadian Radio and Television Commission, which handed out broadcast licences and also went on to set quotas for Canadian content on radio and television. Another was the Canada Council for the Arts. Although founded in 1957, it only began to receive annual appropriations from the federal government in the late 1960s. Both nurtured homegrown talent that might never have gotten a hearing in a world where publishing and music production was financed solely by private for-profit entities. The Canadian Broadcasting Corporation and its French-language wing, Société Radio-Canada, were founded even earlier, but, particularly after the CRTC set Canadian-content quotas, provided information, arts, and music with a distinctly Canadian slant. The right to set CanCon requirements was a major concern of cultural industries during the negotiations over the North American Free Trade Agreement, which led to the Canada-United States-Mexico Agreement in 2018; despite American objections it appears that much of the "cultural exemption" was retained.[68]

A case can be made that the CBC/Radio-Canada has also guaranteed that Canadian media haven't been hijacked by right-wing voices, as is the case in the United States: there is no equivalent in Canada of Fox News or the iHeart-Media network of radio stations.[69] However, the scrutiny that an independent broadcaster can bring to bear on those in power has been well understood by various Canadian governments looking for a place to cut budgets. The Liberals under Jean Chrétien repeatedly cut funding to the CBC, beginning with a $400 million reduction between 1994 and 1997, and Harper and his Conservatives continued the decline while making sure that the corporation's top managers and board were lukewarm supporters of public broadcasting, at best.[70] Justin Trudeau's Liberals, on the other hand, promised an increase of $675 million over five years, and appointed a new head who is viewed as being committed to public broadcasting. What happens next remains to be seen...

When it comes to promoting culture, the provinces have also gotten into the act. Quebec led the way with a concerted effort to set up a star system in television and films, and it has succeeded amazingly well. Part of this is due to the linguistic isolation of the market, some say, but there's more to it than that. Giving a lot of exposure to local artists has encouraged them to dream greater and go farther. Megastar Céline Dion, for example, was once just a little girl singing in bars in a Montreal suburb, but she might never have made it as big as she has without all that CanCon air time at the beginning of her career, particularly once she began singing in English.

South of the border, it goes without saying that there are no quotas pushing American content on broadcasters. Good data on public and private support for the arts is hard to find—the most complete study dates to the late 1990s—but the trend is clear: governments in the United States spend considerably less on culture per capita than do governments in Canada or Europe.[71]

— — —

Given these efforts to reinforce a distinctly Canadian national identity, how to explain the fact that Stephen Harper and his Conservatives won three elections in a row, producing two minority governments and one majority? Does this mean that political differences between the United States and Canada are exaggerated?

No. First of all, Harper, for all his right-wing ideas, was never as bombastic and jingoistic as Donald Trump or the other Republican hopefuls whom Trump defeated to become the GOP's standard-bearer in 2016. After Harper stepped down as party leader in 2015, one candidate for the Conservative leadership, Kellie Leitch, was as xenophobic as Trump is, but she got nowhere in the leadership race held in 2017.[72] A year later, when avowedly populist Doug Ford won the leadership of the Ontario Progressive Conservatives and then a provincial election, his platform, significantly, wasn't anti-immigration, even though it was hard to the right on many other issues.

Even in Quebec, where a few months after Ford's victory another right-wing party, the Coalition Avenir Québec won a provincial election after calling for a decrease in immigration and advocating a "charter of Quebec values," party leader François Legault was adamant that immigrants were wanted. They just have to have the right skills and not be so numerous that they overtax agencies charged with integrating them into Quebec society, he said.[73] While both Legault and Ford shouted quite a bit after their elections about the need to stem *illegal* immigration, their stump speeches pale in comparison to Trump's.

Furthermore, on the federal level, Harper's victories never represented the political ideas of a majority of Canadian voters, not even in 2011, when they sent a Conservative majority to the House of Commons.

In fact, let's stop here for a moment to look more closely at that assertion, in order to note the way the American and Canadian systems both frequently fail to translate the popular vote into governments that accurately reflect the majority's wishes. In Canada voters elect members to the federal House of Commons and to provincial legislative assemblies based on districts or ridings

of more or less equal population size, and in which at least three parties usually run serious candidates. Governments are formed by the party holding the
most seats, with the party's leader becoming the prime minister or the provincial premier, but a huge win in one riding ends up carrying the same weight as
a razor-thin majority in another. This first-past-the-post system is currently
much criticized, and yet, while various forms of proportional representation
have been proposed to replace it, it is a long way from being changed.

In the United States, presidents are elected not by popular vote but by the
538-member Electoral College, a hangover from the late eighteenth century
in which each state was allotted members in proportion to the number of
its representatives and senators. Each state has two senators no matter its
population. They are elected for six-year terms on a rotating basis, so that
every two years some Senate seats come up for election. This can lead to some
curious outcomes in a country where only two parties are real contenders—
as in 2018, when Democrats won 57.5 percent of the popular vote in Senate
contests, but Republicans increased their control of that body.[74] In the House
of Representatives all seats—which are supposedly allotted on the basis of
population—come up for election every two years, but state legislatures can
and do manipulate the boundaries to the advantage of one party or the other.

The Canadian system played in Stephen Harper's favour for three elections
because the other parties—all to the left of his Conservatives—split the opposition vote. In each election, they polled between 8.4 and 8.5 million votes,
compared to the Conservatives 5.3 to 5.8 million. Then in 2015 Harper's streak
of luck ended, and voters for the first time—seemingly independently—
decided that they'd vote for whichever party had the best chance of beating
him. The Cons still polled 5.6 million votes, but Justin Trudeau's Liberals won
6.9 million. Voter turnout was high too, and in the end 11.8 million voted an
anti-Harper ticket, either Liberal, NDP, Green, or Bloc Québécois. That means
that more than two-thirds of Canadian voters cast their ballots against Harper
and his friends.[75] Trudeau and his Liberals are up for re-election in October
2019, when there will be even more parties to split the vote. That could work
in the Libs favour, since competition on the right will be split between the
Conservatives and the new People's Party of Canada, while on the left the NDP
and its leader Jagmeet Singh seem to be struggling.[76] The Liberals were beset
by controversy in the spring of 2019, but, nevertheless, it is almost certain
that a majority of Canadians will vote for a centre-left or left-wing party in
October. Whether that translates into a Liberal majority remains to be seen
at this writing.

In the 2018 Ontario and Quebec provincial elections, centre-left parties also received more than the majority of votes, but the vote split handed wins to centre-right and right-wing parties. In Ontario the provincial Progressive Conservatives won 40.49 percent of the popular vote, with the NDP polling 33.57 percent and the Liberals 19.59 percent.[77] A similar thing happened in Quebec, where the Coalition Avenir Québec won only 37.42 percent of the popular vote. The more centrist Liberals and Parti Québécois, along with the avowedly social democratic Québec Solidaire, split the rest.[78]

In the United States, however, the Right-Left split is much closer and more bitter, reflecting, observers say, a much more polarized population and the greater influence of right-of-centre ideas. Since 2000 the popular vote has gone Democrat in every presidential election except the 2004 race, and in 2016 Hillary Clinton won the popular vote bigly (as Trump would say), with 48.5 percent to Trump's 46.4 percent. Yet even though that difference amounted to nearly three million votes, she still lost in the Electoral College.[79]

Two years later, in the 2018 mid-term elections, voters turned even more to the Democrats. Their congressional candidates scored 53.4 percent of the vote to the Republicans' 44.8 percent. As noted, Democratic Senate candidates did even better in the popular vote, but because of the way the vote was distributed geographically, the Republicans picked up two seats in the Senate.[80]

Dismay over the election results in November 2016 led, famously, to the crash of the Government of Canada's immigration website on election night: it seems it couldn't handle the traffic generated by Americans who were exploring their options before Donald Trump took office.[81] Over the next two years there turned out to be only a small uptick in the number of Americans making the big move, but this interest brings into focus the role that immigration has played in making and keeping Canada and the United States so different.[82] It also underlines the importance of selective immigration.

— — —

The cold water that surrounds Canada on three sides—the Pacific, Atlantic, and Arctic Oceans—has insulated the country to some extent from uncontrolled immigration. So has the fact that to reach it from the south, you have to traverse the United States, a country whose immigration policy has for a long time been in sharp conflict with the Statue of Liberty's promise of an American haven for "your tired, your poor, / Your huddled masses yearning to breathe free." The truth is that, since the middle of the nineteenth

century, the United States has never welcomed as many immigrants proportionately as Canada has.

This is not to say that Canada has greeted everyone enthusiastically. From
the beginning of the country until after World War II those who gained entry
were either recruited for particular ends, or were thought to be "people
like us."[83] As mentioned earlier, many British nationals who fled the upstart
United States during and after the American Revolution were offered land
in Lower and Upper Canada (now Quebec and Ontario). At the end of the
nineteenth century, political and commercial interests combined to recruit
Europeans to settle the great expanses of the Prairies. Between 1874 and 1880,
eight thousand Russian Mennonites settled in Saskatchewan, for example.
Often these immigrants received help from foreign benefactors, particularly
when they faced persecution at home. The foundation set up by the Jewish
financier Baron Maurice de Hirsch bankrolled several farming settlements of
Russian Jews on the Canadian Prairies during the late nineteenth and early
twentieth centuries, while Russian novelist Leo Tolstoy financed the immigration of seventy-five hundred members of the dissident Doukhobor sect with
sales of his novel *Resurrection*. The result of immigration of all sorts can be
seen in those colour-coded maps based on the 1911 census we mentioned earlier, which showed the ethnic origins of people in Alberta and Saskatchewan,
and so looked like a patchwork quilt. Montreal, Toronto, and Winnipeg also
became home to thousands of immigrants from abroad.

At the time of the 1911 census, about 20 percent of all Canadian residents
were foreign born, and the percentage held relatively steady into the 1930s.
But then a combination of the demise of older generations of immigrants and
the almost complete suppression of immigration following World War I saw
the percentage plummet for reasons that were in part racist, part economic.
A governmental Order-in-Council in 1923 banned immigrants "of any Asiatic
race," the definition of which included Armenians seeking refuge from persecution in Turkey. Then, because of anti-Semitism, Canada admitted only five
thousand Jewish refugees between 1933 and 1945.

That was shameful, but note carefully: at its lowest point, in 1951, the proportion of foreign-born Canadians was just a shade under 15 percent of the
overall population, a figure that corresponds with the *highest* point of immigration in US history, 1890.[84]

And therein lies one of the major differences between the two countries
today. There are more immigrants proportionately in Canada, and there have
been for a long time. The country began recruiting immigrants again after World

War II, signing up men and women in European displaced persons camps to work in garment factories, in mines, and on farms. By the 1950s immigration agents were visiting the Azores, nine of Portugal's Atlantic islands, and taking only men who had "hardened hands," because what was wanted was manual labourers: no lawyers need apply. A program to loan European immigrants the cost of passage was also begun in 1951; the money had to be paid back in two years. At first, few immigrants came from non-European countries, but in the 1960s that changed dramatically. By 2011 the two ethnic origins reported most often by first-generation Canadians were Chinese and East Indian, and once again one in five Canadians had been born elsewhere, with the share rising to 27 percent in Ontario. Even in Quebec—which has favoured French-speaking immigrants for two decades, thus reducing considerably the pool from which it recruits—12.9 percent of the population is foreign born, not much under the percentage (13.5) of the United States as a whole in 2016.

So what does this mean?

That right-wing "identity politics"—that is, attacking immigration and foreigners in order to rally support for right-wing parties—will not get a politician nearly as far in Canada, where so many people are newcomers, as it can in the United States. It is true that would-be asylum seekers who have been crossing into Canada from the United States at unpoliced spots along the border don't get much sympathy, and both Ontario's Doug Ford and Quebec's François Legault have angrily demanded the federal government pay for the care and processing of these people. But the immigrants that Canada chooses to admit are a different story, and the general plan is to bring in a lot more—specifically 330,000 new permanent residents in 2019, and 340,000 in 2020.[85]

The operative word here is "choose," which is what Canada actively decided to do back in 1967. Since then it has used a point system to screen those who want to immigrate to the country. When we arrived, I remember being very pleased that I had enough points for my education, training, language, age, and other things to be allowed in on my own steam, and didn't have to rely on my husband's credentials to become a landed immigrant (or what is now called simply a permanent resident.). That year, Canada's foreign-born population amounted to about 16 percent of the total populace, and we were among 184,000 landed immigrants accepted.[86]

Things were very different in the United States at that point. There had been very little choosing on objective criteria, and immigrants to the country were almost exclusively from Europe because 70 percent of immigration quotas from the 1920s to the mid-1960s had gone to people from "old stock" countries like Ireland and the United Kingdom. In 1970, two years after we came to Canada, the census showed that the percentage of the US population born abroad was at its historic low of 4.7 percent. Reforms begun in 1965 were changing things somewhat, since they allotted 6 percent of visas each year to refugees, 74 percent for family reunification, 10 percent for "professionals, scientists and artists," and 10 percent for workers in short supply in the country. The problem of diversity among immigrants was addressed directly in 1990 when a lottery allowing 50,000 immigrant visas was set up, but that's a drop in the bucket when compared with the total number of immigrant visas given out each year: in 2015 some 1,051,031 were.[87]

That's legal immigration; the illegal sort is another matter. Until well into the twentieth century Americans and Canadians went back and forth across a border that was practically without barriers. Wallace Stegner's family, mentioned in the previous chapter, certainly had no official papers when they moved to Wolf Willow. Nor did my Grandfather McDonald and his family, who made two cross-border journeys, nor my Grandfather McGowan, who went from the Prairies to Washington State about the turn of the twentieth century, and may have crossed the Medicine Line often when he was knocking around as a young man. Technically speaking, they were "illegals," but nobody stopped them.[88]

That was about the time that thousands of Mexicans were "invited" to work in the fields of California, Texas, and other border states. Many were shipped back when demand for their labour slackened during the Great Depression, but when demand shot up again during World War II, Mexican workers were again welcomed. The Bracero Program was only supposed to be a wartime measure, but it lasted twenty-two years, and under it 4.6 million Mexicans entered the United States legally as temporary agricultural workers.[89] In addition, however, agricultural workers not covered by the program slipped illegally across the border. The two streams of migrant labour combined to set a pattern of Mexican workers crossing the border to work in the United States for a time. Most of them returned to Mexico eventually, but by 1986 the number of illegal immigrants living in the United States had grown so large that 2.7 million received amnesty when US immigration law underwent another partial reform. The new immigration regulations came with measures that shifted the onus to employers to check if those working for them were legally in the country, but that did not stem the flow. In 2015, an estimated 11.5 million illegal immigrants were in the country, and they were so important to the economy that the business news service Bloomberg warned that the country would have trouble feeding itself unless the visa system was revised to make it easier for agricultural workers to enter legally.[90]

— — —

What will the United States do about immigration besides brandish the threat of criminal hordes storming the border in order to appeal to xenophobes? It's true that in his first State of the Union address, Donald Trump himself suggested that US immigration policy be reformed along Canadian lines by the installation of a "merit-based" system.[91] On election night 2018 his press secretary, Sarah Huckabee Sanders, mentioned doing so in her comments too.[92] But in the run-up to the mid-term elections, Trumpians delivered a message full of anti-foreigner, anti-immigration, anti-anyone-who-isn't-like-us messages. The spectre raised by Trump and his supporters of all those non-white people "breeding" capitalizes on negative attitudes toward people of colour that go back to the country's history of slavery, and it resonates deeply with a segment of the population.[93]

However—and this is a prospect that warms my heart—bad-mouthing immigrants and people of colour in general is going to be a losing tactic in the long term, because, whether or not Trump and his cronies like it, the demographic balance in the United States is changing, and that will have serious political consequences. As Derek Thompson wrote in 2018 in the *Atlantic*:

A majority of babies born in the U.S. for the last four years have been non-white. Historic ethnic diversity is not a future the U.S. can choose to accept or reject; it's the only future on its way. And it's a world where Republicans might finally choose to imitate Canadian conservatives by looking to steal immigrants' votes, rather than their children.

He goes on:

International research on xenophobia has found that whites who don't know many foreign-born people are more likely to fear their presence, while those who actually know immigrants are much more likely to have positive attitudes toward them. This is true even in the U.S., where, despite Trump's election, immigration is more popular than any period in the last 30 years.[94]

The results of the 2018 mid-term elections likewise point to large changes in American politics, and practices. Even in California's Orange County, which had been legendarily Republican, four previously Republican congressional districts flipped to the Democrats.[95] This is particularly significant because, as novelist and academic Viet Thanh Nguyen noted in an article for the *New York Times*, Orange County is home to the largest number of people of Vietnamese descent outside of Vietnam, and most Vietnamese immigrants and their children appear to opt for conservative causes in the United States.[96] The subtext is: identity politics can work many ways.

It is also an example of the long shadow that the Vietnam War still casts today.

— — —

As I said earlier, like my husband and me, most of the Americans who came to Canada in the 1960s and '70s—and we numbered more than forty thousand, the largest mass migration of Americans since Loyalists fled after the American Revolution—were dissatisfied with the war in Vietnam.[97] On the other side of the world, that conflict was displacing several million Vietnamese. Once the war finally ended, several countries admitted sizeable numbers of them, with the United States leading the way. Many of the Vietnamese who went to the States had been actively involved in the American-led war effort. Some of them, like the Hmong people, were essentially airlifted out of a tribal setting that had itself been upset by the war, and then deposited in American towns that were unprepared for them. Others owed their admission to close connections with the departing US forces.

Canada, which hadn't been involved in the war, nevertheless admitted nearly as many Vietnamese refugees proportionally as did the United States. A certain amount of selection occurred, both by Canadian officials and by the refugees themselves.[98] Those Vietnamese who came to Canada were less likely to have been involved on the American side, and more likely to have some knowledge of French: having that language skill gave more points in the immigration evaluation.[99] French-speaking Vietnamese had been studying in Quebec for years too, and there already was an embryonic welcoming group there. The result was a rapid integration of Vietnamese into Canadian society, particularly in Quebec, and, unlike in the United States, a number of young Vietnamese Canadians have become involved in left-wing politics, also particularly in Quebec. Two—Hoang Mai and Anne Minh-Thu Quach, the MPs for Brossard-La Prairie and Salaberry-Suroît, respectively—were elected to the House of Commons in 2011 when the NDP, Tommy Douglas's party, won enough seats to form the Official Opposition. In addition, Ève-Mary Thaï Thi Lac was an MP from Saint-Hyacinthe–Bagot for the leftish Bloc Québécois from 2007 to 2011.

Self-selection has also played a role in population movements at other points in Canadian history. A recent example is the country's successful campaign for universal, single-payer health care, which had the effect of chasing medical professionals who didn't like "socialized medicine" south across the border.[100] More recently, however, the Canadian system has attracted physicians *to* Canada who are fed up with the organization of health care in the United States.[101]

In other words, a significant number of people who don't like what's going on will leave. Like the Carthaginians encouraging malcontents to colonize elsewhere, or like Fidel Castro allowing opponents to leave Cuba in the 1980s, letting people go who don't like what is going on reinforces the dominant values of a place.

— — —

Which brings us to Montreal—specifically to the Feast of the Ascension of Mary on a hot August Sunday, and what might count as a happy ending.

More than four decades had passed since we arrived in Montreal on another sultry late-summer day. This time, instead of getting our first glimpse of McGill, our five-year-old granddaughter Jeanne was taking us to show off the school where she'd start kindergarten in a week or two. The neighbourhood was one that had been built in the early twentieth century. Its modest duplexes and triplexes had been gentrified somewhat, but the neighbourhood still retained elements of the dense, pedestrian-friendly urban village of yore.

The school itself was approaching its hundredth birthday, and was bursting at its seams with the progeny of the young families who'd chosen the neighbourhood for its many attributes. On the corner was Église St-Marc, a Roman Catholic church built about the same time as the school, with a capacity of fifteen hundred; it had seen generations of French Canadians be baptized, confirmed, married, and buried. But one of the things that went hand in hand with Quebec's Quiet Revolution was the secularization of its society, and a precipitous drop in the number of practising Catholics.[102]

At the same time, though, the Vietnamese Catholic community in Montreal had been growing. The church that the parish of Les Saints Martyrs du Vietnam—named for those priests and simple communicants mentioned in the chapter on Vietnam whose execution was the justification for the initial French involvement in Indochina—had been using for a number of years was now too small to hold the thousand or more who regularly attended Sunday services in Vietnamese. Église St-Marc, in contrast, was nearly always empty for French-language services. It had therefore been proposed that the Vietnamese parish use Église St-Marc, and this seemed like a win-win situation for a church that was threatened with closure. There'd be a mass in French on Sundays at 9 a.m. and afterwards religious activity could continue in Vietnamese.[103]

Coming full circle, from Vietnam to Montreal. Vietnamese immigrants of the parish of Les Saints Martyrs du Vietnam have found a spiritual home in Église St-Marc in Montreal, where mass is held in French at 9 a.m. on Sundays, and in Vietnamese the rest of the time. *Photo provided by author.*

That Sunday was the first grand celebration in the church for the Vietnamese congregation. As Jeanne led us down the block to check out the school we saw that the church doors were open, and the sound of many voices singing Vietnamese hymns came floating outside. We walked around the school, cutting through the playground, where Jeanne tried out the climbing equipment. Then we walked back up the next street, passing the church on its other side. The side doors were open, and the music was louder. The singing had stopped, replaced by the haunting tones of what I guessed were Vietnamese instruments. Out in front a couple of young mothers in traditional Vietnamese dress pushed strollers on the sidewalk, while a few small children dressed in their proverbial Sunday best ran up and down the granite steps.

Always curious, my daughter and I watched from a distance for a while, but then ventured up to take a closer look: our menfolk and Jeanne were too shy or too embarrassed to join us. What they missed was the sight of the church full of smiling people watching a group of children doing a graceful dance at the front near the chancel.

A young clergyman, apparently of Vietnamese origin, came over to greet us by the door. We were welcome to enter, he said in French, but we stepped back. It was an event that belonged to a group who had overcome great odds to make a new place for itself in the world, and another chapter in the way places change.

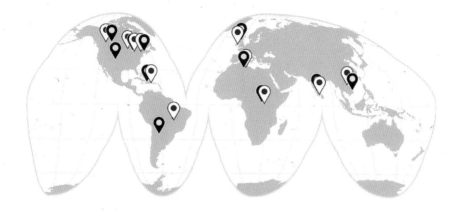

TWINS:
THE LAST WORDS

For a long time, when I reflected on the differences between places that have a lot in common, I thought about them as unidentical twins. Indeed, that's what I wanted to call this book, and several times, as you may have noticed, I've referred to pairs of places that are alike but not alike in those terms. But my publishers said no: a title like that would relegate it to the parenting section of book stores. And yet it seems to me that the comparison is apt. Identical twins occur at a rather steady rate around the world, about three in every thousand births.[1] They are, in effect, clones: after fertilization, the egg-that-will-become-a-person divides completely and almost immediately, to go on to grow into two individuals with exactly the same genetic makeup.

At first glance, the comparison seems to have little to do with a serious discussion of why one political entity is different from a neighbouring one. But, just as what happens after twins are conceived shapes how each individual develops, so the unfolding history of a place affects what it becomes. I am reminded of two girls who were high school friends of mine long before the thought of leaving the United States occurred to me. They were identical twins—beautiful, athletic, funny, and very hard to tell apart. But their paths diverged. A couple of marriages, children (or the lack of them), different career choices, various health problems, and now they don't look much alike at all. Their approach to life is different too: one has always been more flamboyant and impulsive, while the other trundled along with her eye on the main chance. I don't know if either has regrets, but I do know that they've come a long way from the days when the only sure way to tell them apart was to look at their feet: one had a left foot bigger than her right, perhaps because when her embryonic self was separated from that of her sister, a tiny, tiny mistake was made in the division of their DNA, or because as they developed in utero her foot was somehow cramped.

So if two people who are genetically the same can turn out so different, we should not be surprised when similar places develop differently. What is of more interest is what factors account for the divergence. Determining them may teach us something about how to make a better, more equitable world.

At the beginning of this reflection, I suggested that these factors might include geographic variations, no matter how minimal; a colonial history that sometimes depends on a toss of the dice half a world away; exposure to the wider world; the traditional place awarded women; the shelter that a language can give; how the people are educated; and the currents of population movement, including self-selection by the population.

The sample considered here is so small and chosen in such an idiosyncratic manner that blanket assertions about lessons to be applied elsewhere would be foolish. But it seems to me that some things still hold as general principles. There are "givens," like geography, just as human twins have a supremely important "given"—their genetic makeup.

What happens to those "givens" is up to the people who live in the territory concerned. In all cases the people of a political entity have minimal control over many factors, but that has not stopped them from frequently trying to write their own history. The two best examples here, I think, are Haiti's rebellion against France, which was magnificent in its audacity, even though the combined forces of French bankers and American delusions of grandeur set the stage for failure, and Kerala, which took a leftward path after India gained independence, and succeeded spectacularly in improving its citizens' living conditions. Tamil Nadu has also done well, but not as well.

Concerted effort to educate the population made a difference in Kerala, as it has elsewhere. One reason for the differences between Ireland and Scotland lies in the way the Scots—albeit almost inadvertently—set the scene for the

Industrial Revolution by educating their entire population. Similarly, the resources Tunisia poured into education has made a terrific difference in producing a population that seems to be better equipped than its neighbours to govern itself wisely.

Efforts to maintain a native language in the face of colonial pressures also make a difference, as seen in the stories of Vietnam and the two Dravidian twins. The corollary of these efforts—the conscious nurturing of a cultural identity—is responsible for Canada's success in maintaining itself in the face of the American giant. More than once we've also seen what happens when peoples have different traditions governing the place of women: usually giving women a stronger role has gone hand in hand with making a stronger polity.

Lastly, there's the whole matter of population movements, whether by self-selection, forced migration, or in response to economic opportunities. From the Carthaginians to Castro, from the United Empire Loyalists to Canadian Doctors for Medicare, the choices individuals make when faced with divergent opportunities frequently reinforce the currents within a polity.

Change is inevitable, but sometimes it can be channelled, and sometimes a people are lucky. Sometimes the twins are both strong despite—or because—of their differences. Sometimes not, and that is food for thought.

AUTHOR'S **NOTE**

I once heard a famous Canadian writer (I think it was Margaret Atwood) explain the importance of editors: among other things, an editor makes sure the writer doesn't have spinach caught between her teeth, doesn't make silly mistakes. Therefore I'd like to thank an anonymous reviewer of this book who pointed out a few really stupid ones I'd made. Thanks also to Sean Prpick and Bruce Walsh of the University of Regina Press for pertinent criticism and suggestions. Ryan Perks also caught a couple of blunders I shouldn't have made. Of course, any mistakes that remain are my fault and I am responsible for the opinions expressed herein.

NOTES

1: THAT TRIP TO MONTREAL

1   We obviously were not alone. See Penney Kome's *Granny Was a Draft Dodger* (forthcoming from the University of Alberta Press) and Rita Deverell's *American Refugees: Turning to Canada for Freedom* (University of Regina Press, 2019) for more about US immigration to Canada during the 1960s and '70s.

2: THE TWO VIETNAMS

1   If you'd like to take a virtual tour of the pass, there are many, many videos of the trip on YouTube. See, for example, the one made in 2008 on the BBC series *Top Gear*, available at https://www.youtube.com/watch?v=O1zfuBgCUqY, and a more recent one from a young American, Jordan Pike, available at https://www.youtube.com/watch?v=-_LViamSLKQ. The difference between Pike's experience and those of American soldiers is instructive.

2   "Original vestige unearthed at the Hải Vân Gate," *Vietnam Net*, August 27, 2018, https://english.vietnamnet.vn/fms/art-entertainment/207504/original-vestige-unearthed-at-the-hai-van-gate.html/.

3   For an example, see "HAI VAN PASS" at http://huehoianprivatecar.vn/travel-tips/hai-van-pass.html.

4   See Michael Fry, "National Geographic, Korea, and the 38th Parallel: How a National Geographic map helped divide Korea," *National Geographic*, August 4, 2013, https://news.nationalgeographic.com/news/2013/08/130805-korean-war-dmz-armistice-38-parallel-geography/.

5   See "The 1954 Geneva Conference" (documents prepared for New Evidence on the 1954 Geneva Conference on Indochina, Wilson Center, Washington, DC, February 17–18, 2006), https://www.wilsoncenter.org/publication/the-1954-geneva-conference.

6   See reminiscences by Robert Flynn in the University of Arizona's Sixties Project, available at http://www2.iath.virginia.edu/sixties/HTML_docs/Texts/Narrative/Flynn_Street_Joy.html.

7   Bernard B. Fall, *Street Without Joy* (Harrisburg, PA: The Stackpole Company, 1967), 144–73.

8   When I stumbled on these films I wasn't sure if they were authentic. But further investigation shows that they are. Short clips from them were used in official US Army films. The films were shot by SP/4 Walker of the US Army's 1st Infantry Div., 2nd BN, 18th Infantry, on August 30 and 31 and September 1, 1968. The raw stock is now archived at *Critical Past*, available online at https://www.criticalpast.com/video/65675040367_UH-1D-helicopter_Huey_soldiers-cross-jungle-stream_Vietnam-war.

9   Sung by folksinger and anti-war activist Pete Seeger, the song actually referred to an incident in 1942, when US soldiers training in Louisiana were ordered to march into a swamp by their commanding officer, who drowns because he doesn't know the water is deeper than it appears. When Seeger sang it on the popular Smothers Brothers comedy show in 1967, the segment was censored. See "Vietnam Music Monday: 'Waist Deep in the Big Muddy,'" *Newseum*, August 3, 2015, http://www.newseum.org/2015/08/03/vietnam-music-monday-waist-deep-in-the-big-muddy/.

10 BCE (formerly BC) refers to Before Common Era, and CE (formerly AD), to Common Era.

11 Much of this discussion of Vietnamese history is informed by K.W. Taylor's excellent (and long) *A History of the Vietnamese* (Cambridge: Cambridge University Press, 2013). I have not given individual page notes, but for more detail, see in particular chapter 1, "The Provincial Era," 14–50; chapter 4, "The Le Dynasty," 165–223; chapter 9, "The Nguyen Dynasty," 398–445; chapter 12, "Indochina at War," 524–60; and chapter 13, "From Two Countries to One," 561–606.

12 Mark Stoneking and Frederick Delfin, "The Human Genetic History of East Asia: Weaving a Complex Tapestry," *Current Biology* 20, no. 4 (2010), https://doi.org/10.1016/j.cub.2009.11.052. More evidence of the complex migration that took place in the region is turning up in DNA analysis of skeletons from four to eight thousand years ago. See Hugh McColl et al., "The Prehistoric Peopling of Southeast Asia," *Science* 361, no. 6397 (2018), https://doi.org/10.1126/science.aat3628.

13   Mark J. Alves, "What's so Chinese about Vietnamese?" in *Papers from the Ninth Annual Meeting of the Southeast Asian Linguistics Society*, ed. Graham Thurgood, 221–42 (Tempe: Arizona State University, Program for Southeast Asian Studies, 1999).

14   Heather Peters, "Tattooed Faces and Stilt Houses: Who Were the Ancient Yue?" *Sino-Platonic Papers* 17 (1990), http://sino-platonic.org/complete/spp017_yue.pdf.

15   Taylor, *History of the Vietnamese*, 15.

16   See "Nam Viet," *Encyclopaedia Britannica* (2007), https://www.britannica.com/place/Nam-Viet.

17   Taylor, *History of the Vietnamese*, 16.

18   See Pamela LaBorde, "Vietnamese Cultural Profile," EthnoMed, December 2010, https://ethnomed.org/culture/vietnamese/vietnamese-cultural-profile; and Jeffrey Hayes, "Women in Vietnam: Traditional Views, Advances and Abuse," *Facts and Details* (blog), May 2014, http://factsanddetails.com/southeast-asia/Vietnam/sub5_9c/entry-3390.html#chapter-2Vietnamese.

19   See former diplomat Steven B. Young's series of six lectures on Vietnamese history delivered in 2014 under the auspices of the Vietnamese Center of Minnesota, and available at https://www.youtube.com/watch?v=T9yg9uvosxu. Young's analysis is informed by his own experience in Vietnam, and has a pro-American point of view, but the series is a valuable introduction to the subject, particularly when taken in combination with Taylor's masterful *History of the Vietnamese*.

20   Fall, *Street Without Joy*, 284.

21   Marco Polo, *The Travels*, trans. Ronald Latham (London: Penguin Classics, 1958), 249.

22   The Mỹ Sơn temple complex is a UNESCO World Heritage Site. For information about it, and a short appreciation of the Cham civilization, see UNESCO World Heritage Centre, "My Son Sanctuary," http://whc.unesco.org/en/list/949/.

23   The planes had a wingspan of 56.4 metres (185 feet) and a length of 48.5 metres (or 159 feet). They could carry a payload of 31,500 kilograms (70,000 pounds). See The Boeing Company, "B-52 Technical Specifications," http://www.boeing.com/defense/b-52-bomber/#/technical-specifications.

24   For examples, see https://www.youtube.com/watch?v=88jrZjsNHPc and https://www.youtube.com/watch?v=Ba87xDCqHaQ.

25   A third system, Taoism, also entered Vietnam under Chinese rule. Its belief that the universe is in constant tension between *yin* and *yang*, the female and male principles, led to complex rituals and a large pantheon of gods and immortals presided over by the Jade Emperor. Note also that some scholars push back Zoroaster's dates by nine hundred years or more, based on linguistic analysis of the founding text, the Yasna. See "Zoroaster," *New World Encyclopedia*, 2013, http://www.newworldencyclopedia.org/entry/Zoroaster.

26   It should be mentioned that the Chams were not the only people living in the southern Indochinese Peninsula. In addition to fiercely independent people in the

highlands, another group controlled the Mekong Delta, and, like the Chams, were intimately connected to Indian civilization. The kingdom of Funan—the name given them by the Chinese (we do not know what they called themselves)—thrived between the first and sixth centuries CE. Archeological evidence shows that they had contact with China, Persia, India, and even the Mediterranean. But before the Viets succeeded in shaking off Chinese domination, they had been absorbed into the kingdom of the Khmers in what is now Cambodia. See Georges Coedès, *Les Peuples de la Péninsule Indochinoise: Histoire—Civilisations* (Paris: Dunod, 1962), 62.

27 Some Chams remained in small pockets in what is now Vietnam, while others moved into what is now Cambodia.

28 See Joachim Pham, "Jesuits launch jubilee year to mark 400 years in Vietnam," *National Catholic Reporter*, February 10, 2014, https://www.ncronline.org/news/ parish/jesuits-launch-jubilee-year-mark-400-years-vietnam.

29 Taylor, *History of the Vietnamese*, 288.

30 Ibid., 322.

31 See "Martyrs of Vietnam, *Catholic Online*, 2019, https://www.catholic.org/saints/ saint.php?saint_id=4951.

32 See Taylor, *History of the Vietnamese*, 468–73.

33 Paul Theroux, *Ghost Train to the Eastern Star: On the Tracks of the Great Railway Bazaar* (Boston and New York: Houghton Mifflin, 2008), 387.

34 John Springhall, "'Kicking out the Vietminh': How Britain Allowed France to Reoccupy South Indochina, 1945–46," *Journal of Contemporary History* 40, no. 1 (2005): 115–30.

35 Ho Chi Minh's speech can be read at http://historymatters.gmu.edu/d/5139/.

36 For a more detailed account of why the Soviets first became involved in the Vietnamese conflict and then tarried before pulling out, see the Sino-Soviet expert Sergey Radchenko's analysis in "Why Were the Russians in Vietnam?" *New York Times*, March 27, 2018, https://www.nytimes.com/2018/03/27/opinion/ russians-vietnam-war.html. Radchenko is a professor of international relations at Cardiff University, Wales, and his article is part of a series reflecting on the fiftieth anniversary of the American war in Vietnam. See also Taylor, *History of the Vietnamese*, 605–26.

37 Reunification Day (April 30) is a national holiday marking the fall of Saigon, although the country wasn't fully reunited until July 2, 1976, when the Socialist Republic of Vietnam came into being.

38 "How Tet festival different between the Northern and the Southern Vietnam?" *Travel Sense Asia*, 2014, http://www.travelsense.asia/ how-tet-festival-different-between-the-northern-and-the-southern-vietnam/.

39 Taylor, *History of the Vietnamese*, 642.

40 "VN among top 10 countries receiving remittances," *Vietnam News Agency*, March 1, 2018, https://vietnamnews.vn/economy/423560/vn-among-top-10-countries- receiving remittances.html#gXrGcu68mHxKe94k.97.

3: ALGERIA **AND TUNISIA**

1   Richard Miles, *Carthage Must be Destroyed: The Rise and Fall of an Ancient Civilization* (New York: Viking, 2010), 58–70.

2   Note that the dictionary definition of "indigenous" is always a variation on "people or things that belong to the country in which they are found, rather than coming there or being brought there from another country" (in the words of the *Collins English Dictionary*, https://www.collinsdictionary.com/dictionary/english/ indigenous). "Indigenous" (with a capital I) has become the preferred name for those people who settled the Americas before the arrival of Europeans. It will be used here in preference over "Native Americans," "First Nations," and "Indians." But "indigenous" (absent capitalization) will also be used to refer to populations existing before the arrival of other peoples, as in North Africa and India.

3   Paul Krugman, *The Conscience of a Liberal* (New York: W.W. Norton & Company, 2007), 3. He writes that this was the result of rising expectations following the great advances in civil society that had brought in Social Security, higher wages, and union protection, as well as such things as a greatly expanded educational system. Much has changed since then.

4   The *New York Times* reported on the riots on October 22, 1961, but that issue of the paper would not have made its way to the West Coast until early November. In the story that the *Times* ran, estimates of the death toll were no higher than twenty. See Robert C. Doty, "Riots Dim Algeria Peace Hopes," *New York Times*, October 22, 1961, available at https://timesmachine.nytimes.com/ timesmachine/1961/10/22/101480579.html?pageNumber=190.

5   See Soren Seelow, "17 octobre 1961: 'Ce massacre a été occulté de la mémoire collective'," *Le Monde*, October 17, 2011, http://www.lemonde.fr/societe/ article/2011/10/17/17-octobre-1961-ce-massacre-a-ete-occulte-de-la-memoire- collective_1586418_3224.html.

6   Jean-Jacques Hublin et al., "New fossils from Jebel Irhoud, Morocco and the pan-African origin of *Homo sapiens*," *Nature* 546, no. 7657 (2017), https://www.nature. com/articles/nature22336.

7   The next-oldest anatomically modern human remains were found in East Africa, and date from about 200,000 years ago. It appears that a small group of them went south to populate sub-Saharan Africa, while another crossed or went around the Red Sea 70,000 to 80,000 years ago. All humans who don't have recent roots in Africa appear to be descended from the latter group, who literally went forth to be fruitful and multiply. See my book *Road Through Time: The Story of Humanity on the Move* (Regina: University of Regina Press, 2017) for more on this topic.

8   Gabriel Camps, "Tableau chronologique de la Préhistoire récente du Nord de l'Afrique," *Bulletin de la Société préhistorique française. Études et travaux* 65, no. 2 (1968): 609–22.

9 See Mary Jackes and David Lubell, "Early and Middle Holocene Environments and Capsian Cultural Change: Evidence from the Télidjène Basin, Eastern Algeria," *African Archaeological Review* 25, no. 1–2 (2008): 41–55, https://doi.org/10.1007/s10437-008-9024-2.

10 D.J. Mattingly and M. Sterry, "The first towns in the central Sahara," *Antiquity* 87, no. 336 (2013): 503–18, https://doi.org/10.1017/S0003598X00049097.

11 David Keys, "Kingdom of the Sands," *Archeology* 57, no. 2 (2004), http://archive.archaeology.org/0403/abstracts/sands.html.

12 See UNESCO, "Archeological Site of Carthage," available at https://www.youtube.com/watch?v=fa_l-d3YoPc.

13 Miles's *Carthage Must be Destroyed* (cited above) presents a well-researched, but wonderfully readable, account of Carthage from its founding until its rebirth as a Roman city.

14 Ben Kiernan, "The First Genocide: Carthage, 146 BC," *Diogenes* 51, no. 3 (2004): 27–39, https://journals.sagepub.com/doi/10.1177/0392192104043648.

15 See Roger D. Woodard, "*Phoinikēia Grammata*: An Alphabet for the Greek Language," in *A Companion to the Ancient Greek Language*, ed. Egbert J. Bakker, 25–46 (Malden, MA: Wiley-Blackwell, 2010).

16 Richard Salomon, "On the Origin of the Early Indian Scripts: A Review Article," *Journal of the American Oriental Society* 115, no. 2 (1995): 271–9.

17 Aristotle, *Politics*, trans. J.E.C. Welldon (New York: Macmillan, c. 1893). Quoted material comes from bk. 2, ch. 11, available at https://archive.org/stream/in.ernet.dli.2015.216306/2015.216306.The-Political_djvu.txt.

18 Alfonso Chardy, "How Fidel Castro and the Mariel boatlift changed lives and changed Miami," *Miami Herald*, November 26, 2016, http://www.miamiherald.com/news/nation-world/world/americas/fidel-castro-en/article117206643.html.

19 A brief sample:
    From the Halls of Montezuma
    To the shores of Tripoli;
    We fight our country's battles
    In the air, on land, and sea;
    First to fight for right and freedom
    And to keep our honor clean;
    We are proud to claim the title
    Of United States Marine
    The song's lyrics are available at http://www.usflag.org/songs.html#MH.

20 The mother of one of the last Roman emperors, Septimius Severus, was from a noble Roman family that had immigrated to North Africa. Severus was himself born in Leptis Magna, a town in what is now Libya that had been a garrison against Berber incursions. See Cassius Dio, *Roman History*, books 74–6, available at http://penelope.uchicago.edu/Thayer/e/roman/texts/cassius_dio/home.html.

21 The film won the Golden Lion at the 1967 Venice Film Festival, but was not shown in France for five years because the French film industry considered it too controversial.

22 Ebert wrote at least two reviews of *The Battle of Algiers*, the first on the occasion of the film's original North American release, and the second after the release of the DVD. The first is available at http://www.rogerebert.com/reviews/the-battle-of-algiers-1968, the second at http://www.rogerebert.com/reviews/great-movie-the-battle-of-algiers-1967. He considered it one of the great movies of all time.

23 Memory is the foundation of my writing, and when I can, I check my memories against other sources—more about that in the chapter on Burundi and Rwanda. In this case, I had seen *The Battle of Algiers* once. Though I'd taken no notes, images from it—particularly of the young woman passing through the checkpoint—were sharp in my memory, and I was able to recall them without any trouble. I found the bombing sequence on YouTube— https://www.youtube.com/watch?v=7hYtN2zWX8c—and was delighted to see that the picture in my head matched the one on the screen. What I had not remembered was that there were three women: an example of my mind stripping things down to the bare essentials, perhaps.

24 Mounira M. Charrad, *States and Women's Rights: The Making of Postcolonial Tunisia, Algeria, and Morocco* (Berkeley and Los Angeles: University of California Press, 2001), 188–99.

25 Charlotte Bozonnet, "En Algérie, 'il reste beaucoup à faire' pour l'égalité des femmes," *Le Monde*, March 13, 2015, http://www.lemonde.fr/afrique/article/2015/03/13/en-algerie-il-reste-beaucoup-a-faire-pour-l-egalite-des-femmes_4593413_3212.html#tmktHUlhhjRJHxD2Y.99.

26 Charrad, *States and Women's Rights*, 230–1.

27 See Harry Ostrer and Karl Skorecki, "The Population Genetics of the Jewish People," *Human Genetics* 132, no. 2 (2013): 119–27, https://doi.org/10.1007/s00439-012-1235-6.

28 John Julius Norwich, *The Middle Sea: A History of the Mediterranean* (London: Vintage, 2007), 72–7.

29 Charrad, *States and Women's Rights*, 22.

30 Quran, Surah An-Nisa 4:11. Mosaic law says a daughter can only inherit from her father if he leaves no sons, and does not mention a widow's right to inherit. Numbers 27: 5–11.

31 Tarek El-Tablawy and Jihen Laghmari, "Tunisia's President Says Women Should Have Equal Inheritance Rights," *Bloomberg*, August 13, 2018, https://www.bloomberg.com/news/articles/2018-08-13/tunisia-backs-equal-inheritance-for-women-in-break-with-region.

32 Charrad, *States and Women's Rights*, 430.

33 At an official reception, the French envoy and the Ottoman lord began by bringing up events that had been rankling their countries for some time. One word led

to another, and by the end of the exchange, both were outraged. The Ottoman ruler capped things off by switching the French diplomat with his fly whisk, an ornamental fan that did little damage physically but the symbolism of which was perceived as an enormous insult. See Norwich, *The Middle Sea*, 497.

34 Beppi Crosariol, "The rise and fall of a wine juggernaut (or why your wine doesn't come from Algeria any more)," *Globe and Mail*, February 26, 2013, http://www.theglobeandmail.com/life/food-and-wine/wine/the-rise-and-fall-of-a-wine-juggernaut-or-why-your-wine-doesnt-come-from-algeria-any-more/article9076845/.

35 Michael J. Willis, *Politics and Power in the Maghreb: Algeria, Tunisia and Morocco from Independence to the Arab Spring* (New York: Columbia University Press, 2012), 22–4.

36 Ibid., 30.

37 Jean-Marc Gonin, "Les pieds-noirs, 50 ans après," *Le Figaro*, October 8, 2012, http://www.lefigaro.fr/actualite-france/2012/01/27/01016-20120127ARTFIG00422-les-pieds-noirs-50-ans-apres.php.

38 See "Algeria," in *The World Factbook* (Washington, DC: Central Intelligence Agency), available at https://www.cia.gov/library/publications/the-world-factbook/geos/ag.html.

39 Willis, *Politics and Power*, 75.

40 Al Jazeera had its official access revoked by the Algerian government in 2004, but in 2012 a film crew was able to make a clandestine documentary about conditions during and after the Arab Spring. See *Algeria: The revolution that never was*, available at http://www.aljazeera.com/programmes/peopleandpower/2012/05/2012516145457232336.html.

41 "Huge protests in Algeria as allies turn on Bouteflika," *Al Jazeera*, March 16, 2019, https://www.aljazeera.com/news/2019/03/fresh-anti-bouteflika-protests-algeria-allies-turn-190315115305675.html.

42 All has not been tranquil in Sidi Bouzid since the Arab Spring, as witness a video of a May 1, 2017, anti-terrorist operation filmed by a Special Forces unit of the Tunisian National Guard. Quite a difference from the "home movies" of the Vietnam War, mentioned in chapter 1. See "Opération sécuritaire à Sidi Bouzid: 2 terroristes abattus, 4 autres arrêtés," *HuffPost Maghreb*, April 30, 2017, http://www.huffpostmaghreb.com/2017/04/30/sidi-bouzid-terrorisme_n_16341216.html.

43 Martine Gozlan, *Tunisie-Algérie-Maroc: la colère des peuples* (Paris: L'archipel, 2011), 17.

44 One such video is available at https://www.youtube.com/watch?v=0x2mdGimsDo.

45 See "Yemen facing largest famine the world has seen for decades, warns UN aid chief," *UN News*, November 9, 2017, https://news.un.org/en/story/2017/11/570262-yemen-facing-largest-famine-world-has-seen-decades-warns-un-aid-chief, and Hannah Summers, "Yemen on brink of 'world's worst famine in 100 years' if war continues," *Guardian*, October 15, 2018, https://www.theguardian.com/global-development/2018/oct/15/yemen-on-brink-worst-famine-100-years-un.

46 Kamel Daoud, "The Algerian Exception," *New York Times*, May 29, 2015, http://www.nytimes.com/2015/05/30/opinion/the-algerian-exception.html?_r=0.

47 Thessa Lageman, "Mohamed Bouazizi: Was the Arab Spring worth dying for?" *Al Jazeera*, January 3, 2016, http://www.aljazeera.com/news/2015/12/mohamed-bouazizi-arab-spring-worth-dying-151228093743375.html.

48 See "Tunisia Predicted: Demography and the Probability of Liberal Democracy in the Greater Middle East," Wilson Center, 2011, https://www.wilsoncenter.org/event/tunisia-predicted-demography-and-the-probability-liberal-democracy-the-greater-middle-east, and Richard Cincotta, "Opening the Demographic Window: Age Structure in Sub-Saharan Africa," Wilson Center, October 26, 2017, https://www.newsecuritybeat.org/2017/10/opening-demographic-window-age-structure-sub-saharan-africa/?utm_source=feedburner&utm_medium=email&utm_campaign=Feed%3A+TheNewSecurityBe.

49 See "Tunisia: Average age of the population from 1950 to 2050 (median age in years)," *Statista*, 2019, https://www.statista.com/statistics/524614/average-age-of-the-population-in-tunisia/. In comparison, Algeria's was just 26, demographer Richard Cincotta's proxy cut-off age for political stability and liberal democracy (see note 48, above).

50 "The Nobel Peace Prize 2015" (press release), August 4, 2017, http://www.nobelprize.org/nobel_prizes/peace/laureates/2015/press.html.

51 Sarah Feuer, "Reshuffle, Rinse, and Reform: Tunisia's Government Under Strain," The Washington Institute, June 12, 2018, https://www.washingtoninstitute.org/fikraforum/view/reshuffle-rinse-and-reform-tunisias-government-under-strain.

52 Reuters New Agency, "Tunisia's powerful UGTT workers union holds nationwide strike," *Al Jazeera*, January 17, 2019, https://www.aljazeera.com/news/2019/01/tunisia-powerful-ugtt-workers-union-holds-nationwide-strike-190117081348601.html.

4: KERALA AND TAMIL NADU

1 Montreal: Véhicule Press, 2006.

2 For a sample, see https://www.youtube.com/watch?v=9QlRlQ7loZU. And it seems that drivers rely on honking more as a safety measure than they do in Vietnam.

3 See India Meteorological Department, "Northeast Monsoon," http://www.imdchennai.gov.in/northeast_monsoon.htm, and Kerala Department of Tourism, "Monsoon in Kerala," https://www.keralatourism.org/monsoon/.

4 Comparisons of DNA from ancient remains and more recent Indians show elements of three populations who met over the centuries and intermarried. One is from hunter-gatherers who apparently colonized the subcontinent tens of thousands of years ago as humankind spread out from its African origins. Another is descended from Middle Eastern farmers, and a third from Central Asian herders.

See Chandana Basu Mallick et al., "The Light Skin Allele of SLC24A5 in South Asians and Europeans Shares Identity by Descent," PLOS *Genet* 9, no. 11 (2013), https://doi.org/10.1371/journal.pgen.1003912, and Lizzie Wade, "South Asians are descended from a mix of farmers, herders, and hunter-gatherers, ancient DNA reveals," *Science Magazine*, April 18, 2018, https://www.sciencemag.org/news/2018/04/south-asians-are-descended-mix-farmers-herders-and-hunter-gatherers-ancient-dna-reveals.

5   Dervla Murphy, *On a Shoestring to Coorg: An Experience of Southern India* (London: Flamingo, 1995), 126.

6   Robert Caldwell, *A comparative grammar of the Dravidian, or, South-Indian family of languages* (London: Trubner & Co., 1875), and Michel Danino, "Genetics and the Aryan Debate," *Puratattva: Bulletin of the Indian Archaeological Society* 36 (2005–6): 146–54.

7   On the relation and evolution of Tamil and Malayalam scripts, see the chart located at http://www.cs.colostate.edu/~malaiya/brah11.gif.

8   See "Sangam Poems Translated by Vaidehi Herbert," available at http://sangamtranslationsbyvaidehi.com/. Used by permission of the translator.

9   See "Kerala Etymology," http://www.stateofkerala.in/kerala_facts/kerala_etymology.php.

10 Strabo, *Geography*, bk. 2, ch. 5, ll. 1–2, available at http://penelope.uchicago.edu/Thayer/e/roman/texts/strabo/2e1*.html.

11 See John Keay, *India: A History* (New York: Harper Collins, 2000); see especially chapter 5 (pp. 78–100) for an account of the Maurya Empire.

12 "Ashoka's Rock Edicts," section 13, *Livius: Articles on Ancient History*, 2004, http://www.livius.org/sources/content/ashoka-s-rock-edicts/.

13 Samar Abbas, "India's Parthian Colony: On the Origin of the Pallava Empire of Dravidia," Circle of Ancient Iranian Studies, http://www.cais-soas.com/CAIS/History/ashkanian/parthian_colony.htm.

14 David Keys, "The lost empire explored: The Cholas once had great power, but the world has forgotten them," *Independent*, May 9, 1993, http://www.independent.co.uk/arts-entertainment/travel-the-lost-empire-explored-the-cholas-once-had-great-power-but-the-world-has-forgotten-them-2321900.html.

15 See "The Periplus of the Erythraean Sea: Travel and Trade in the Indian Ocean by a Merchant of the First Century," sections 53–6, *Ancient History Sourcebook* (Fordham University), October 2000, http://sourcebooks.fordham.edu/ancient/periplus.asp.

16 For a more complete discussion of the role of the Portuguese in this drive to explore the world, see my *Making Waves: The Continuing Portuguese Adventure* (Montreal: Véhicule Press, 2010).

17 Taylor, *History of the Vietnamese*, 272.

18 His bones were shipped back to Portugal some years after his death, I later learned.

19 William Dalrymple, *Doubting Thomas*, film presented on BBC2, April 15, 2000.

20 From the St. Thomas Mount National Shrine website, available at http://www.stthomasmount.org/history.php.

21 The stone boat story, which is part of the great Campostela tradition, might also reflect an earlier pagan burial rite, since "stone boats" were sometimes used in Celtic burials and there is much Celtic tradition in that part of the Iberian Peninsula.

22 2011 Indian census data: http://www.censusindia.gov.in/2011census/C-01.html.

23 See https://www.syriacstudies.com/2013/06/29/st-thomas-christians-the-syrian-merchant-thomas-cana-arrives-in-malabar/.

24 See Kimberley Reine and Boaz Berney's charming account of their visit to the Pardesi Synagogue, where Boaz was warmly greeted as another Jewish man needed to make the minyan (the ten required for Friday night services): "The Last Jews of Cochin," *Lime Soda* (blog), August 23, 2008, http://kimboaz.blogspot.ca/2008/08/last-jews-of-cochin.html.

25 Huizhong Wu, "India's Cheraman mosque: A symbol of religious harmony," *Al Jazeera*, April 23, 2017, https://www.aljazeera.com/indepth/features/2017/04/india-cheraman-mosque-symbol-religious-harmony-170406095923455.html. See also Cheraman Juma Musjid, "History," 2011, http://www.cheramanmosque.com/history.php.

26 "How a Portuguese Jewish Jesuit Produced the First Printed Tamil Book," *Madras Courier*, August 2, 2018, https://madrascourier.com/books-and-films/how-a-portuguese-jewish-jesuit-produced-the-first-printed-tamil-book/.

27 Marco Polo, *Travels*, 274.

28 James Talboys Wheeler, *India Under British Rule from the Foundation of the East India Company* (London: Macmillan, 1886), 41. The edition quoted here comes from Project Gutenberg (2014), and is available at http://www.gutenberg.org/files/46151/46151-h/46151-h.htm.

29 Literacy rates for Kerala are found at http://www.thrissurkerala.com/census-2011/Census%202011-%20Kerala%20Profile.html. For Tamil Nadu education statistics, see http://www.tn.gov.in/deptst/education.pdf.

30 Infant mortality rates are available at https://niti.gov.in/content/infant-mortality-rate-imr-1000-live-births.

31 See https://www.census2011.co.in/census/state/kerala.html and https://www.census2011.co.in/census/state/tamil+nadu.html.

32 See https://www.census2011.co.in/census/state/kerala.html, https://www.census2011.co.in/census/state/tamil+nadu.html, and https://www.census2011.co.in/sexratio.php.

33 Choodie Shivaram, "Where Women Wore the Crown: Kerala's Dissolving Matriarchies Leave a Rich Legacy of Compassionate Family Culture," *Hinduism Today*, February 1996, https://www.hinduismtoday.com/modules/smartsection/item.php?itemid=3569.

34 Bina Agarwal, "Can we unify inheritance law?" *Times of India*, September 19, 2017, https://timesofindia.indiatimes.com/india/can-we-unify-inheritance-law/articleshow/60740547.cms.

35 Snigdha Poonam, "Indian women just did a remarkable thing—they formed a wall of protest," *Guardian*, January 3, 2019, https://www.theguardian.com/commentisfree/2019/jan/03/gender-activism-india-womens-wall-sabarimala-temple-kerala.

36 Palash Ghosh, "Dravida Nadu: What If the South Seceded from the Republic of India?" *International Business Times*, March 3, 2013. http://www.ibtimes.com/dravida-nadu-what-if-south-seceded-republic-india-1413910.

37 Abdul Ruff, "Genocides of Tamils and Indo-Sri Lanka relations," *Modern Diplomacy*, April 3, 2017, https://moderndiplomacy.eu/2017/04/03/genocides-of-tamils-and-indo-sri-lanka-relations/.

38 Whitney Cox, "The Wonderful Allure of Tamil," *New York Review of Books*, March 23, 2017, https://www.nybooks.com/articles/2017/03/23/wonderful-allure-of-tamil/.

39 "Southern comfort: Tamil Nadu and Kerala dance to a different tune from the rest of India," *Economist*, May 28, 2016, https://www.economist.com/asia/2016/05/28/southern-comfort.

40 See "Why Kerala loves its newspaper so much," *News Minute*, June 17, 2018, https://www.thenewsminute.com/article/why-kerala-loves-its-newspaper-so-much-83193.

41 Manu Joseph, "Setting a High Bar for Poverty in India," *New York Times*, July 9, 2014, http://www.nytimes.com/2014/07/10/world/asia/setting-a-high-bar-for-poverty-in-india.html. See also Alexandra Brown, "Growth and Success in Kerala," *Yale Review of International Studies*, November 2013, http://yris.yira.org/essays/1150, and P.V. Rajeev, "Why Kerala has high unemployment," *Market Express*, January 4, 2018, http://www.marketexpress.in/2018/01/why-kerala-has-high-unemployment.html.

42 Rajan Chedambath, interview with author, February 15, 2005.

43 Both Kerala and Tamil Nadu have higher per capita incomes than the Indian average, but both rank lower than individual states like Goa and Delhi. See "Indian states by GDP per capita," available at http://statisticstimes.com/economy/gdp-capita-of-indian-states.php.

5: BRAZIL AND THE REST OF SOUTH AMERICA

1 At that time, the University of California required three years of foreign-language study for admission, and at my excellent public high school in San Diego during the Cold War years, French and Russian were also available. The latter seemed to me to be too hard, and the former too frou-frou, so Spanish it was. My life after we came to Montreal might have been a little easier had I picked French, I've always thought.

2 The finished book is *Road Through Time: The Story of Humanity on the Move*, published in 2017 by the University of Regina Press.

3 Brazil's population was about 212.4 million in early 2019, more than four times as large as the next most populous South American country, Colombia (49.8

million). Peru had 32.7 million, while Mexico, by way of comparison, had 132.3 million. Figures are available at http://worldpopulationreview.com/countries/.

4   Juvenal Milton Engel, "The Brazilian International Boundary Commissions," http://info.lncc.br/cbdlsi.html.

5   David Salisbury, A. William Flores de Melo, and Pedro Tipula Tipula, "Transboundary Political Ecology in the Peru-Brazil Borderlands: Mapping Workshops, Geographic Information, and Socio-Environmental Impacts," *Revista Geográfica* 152 (2012): 105–15, http://scholarship.richmond.edu/cgi/viewcontent.cgi ?article=1032&context=geography-faculty-publications.

6   As mentioned before, my *Making Waves: the Continuing Portuguese Adventure* gives a more detailed look at the adventures and accomplishments of the Portuguese and their descendants around the world.

7   Stanley G. Payne, *A History of Spain and Portugal* (Madison: University of Wisconsin Press, 1973), 1: 28–32. A digital copy can be found at the University of Arkansas's Library of Iberian Resources Online at http:// epicroadtrips.us/2013/ spain/23may2013_Thursday/aHistoryofPortugalandSpain.pdf.

8   For a not-too-detailed account of the *Reconquista*, by which the Moors were expelled from Iberia, see the entry in the *New World Encyclopedia*, available at http://www.newworldencyclopedia.org/entry/Reconquista.

9   This national affirmation came despite the fact that Portugal's population has always been much smaller than that of its neighbour to the east. In the 1400s it was about a million, compared to more than six million in the fragmented territory that would become Spain. See "Population of Western Europe" at http:// www.tacitus.nu/historical-atlas/population/westeurope.htm. See also Payne, *Spain and Portugal*, 120–46.

10 Stephen R. Bown, *1494: How a Family Feud in Medieval Spain Divided the World in Half* (Vancouver: Douglas & McIntyre, 2011).

11 *A Midsummer Night's Dream*, Act 1, Scene 1, 134–6.

12 Bown, *1494*, 148–55.

13 Pope Julius ii, *Ea Quae*, issued January 24, 1506. Christopher Columbus would die within a few months too.

14 For a quick explanation, see "The Portuguese Explorers," Newfoundland and Labrador Heritage Website, August 2004, http://www.heritage.nf.ca/articles/ exploration/portuguese.php.

15 See A. Davies, "Fernandes, João," *Dictionary of Canadian Biography*, vol. 1 (University of Toronto/Université Laval, 2003), http://www.biographi.ca/en/bio. php?id_nbr=212.

16 The linguist Morris Swadesh began his investigation of premodern languages by studying the Salish language, spoken by the Indigenous population of British Columbia. He expanded it from there, and his work was repeated by linguists in many cultures. See Stanley Newman, "Morris Swadesh," *Language* 43, no. 4 (1967): 948–57.

17 Stephen R. Anderson, "How Many Languages Are There in the World?" Linguistic Society of America, 2010, http://www.linguisticsociety.org/content/how-many-languages-are-there-world.

18 Canada's first census, conducted in 1871, recorded that 31 percent of the population were "French." In 1901, the first census in which questions about language were specifically asked, 32 percent were "able to speak French." (See Statistics Canada, "The evolution of English-French bilingualism in Canada from 1901 to 2011," 2018, https://www150.statcan.gc.ca/n1/pub/11-630-x/11-630-x2016001-eng.htm.) In the 2016 census, 21 percent of respondents said French was their mother tongue. See Office of the Commissioner of Official Languages, "Fast figures on Canada's official languages (2016)," https://www.clo-ocol.gc.ca/en/statistics/canada.

19 Henry Kamen, *Spain's Road to Empire: The Making of a World Power, 1492–1763* (London: Penguin, 2009), 3. The following section draws heavily on this book, particularly pp. 301–37.

20 Until recently, Guyane was a French *département*, but in 2016 it and Martinique became *collectivités térritoriales*. Some background on that distinction is available at https://www.vie-publique.fr/decouverte-institutions/institutions/collectivites-territoriales/categories-collectivites-territoriales/que-sont-collectivites-territoriales-martinique-guyane.html.

21 An indication of the length of time a voyage might take comes from that of the *Beagle*, the British ship on which Charles Darwin was resident naturalist. It left Devonport, England, on December 27, 1831, and arrived at Bahia February 29, 1832. In his notes on the voyage, Darwin doesn't mention whether this was good or poor time.

22 Antonio Carlos Pereira Martins, "Ensino superior no Brasil: da descoberta aos dias atuais," *Acta Cirurgica Brasileira* 17, no. 3 (2002), http://dx.doi.org/10.1590/S0102-86502002000900001.

23 John Lynch, "Simón Bolívar and the Spanish Revolutions," *History Today* 33, no. 7 (1983), http://www.historytoday.com/john-lynch/simon-bolivar-and-spanish-revolutions.

24 For a more complete account, see Emilia Viotti da Costa, *The Brazilian Empire: Myths and Histories* (Chapel Hill and London: University of North Carolina Press, 2000).

25 Simon Romero, "Pope's Trip to Brazil Seen as 'Strong Start' in Revitalizing Church," *New York Times*, July 28, 2013, http://www.nytimes.com/2013/07/29/world/americas/vibrant-display-at-popes-last-mass-in-brazil.html.

26 That interview is available at https://www.youtube.com/watch?v=kS2FWCLjSlk.

6: HAITI AND THE DOMINICAN REPUBLIC

1   It was near the end of his return voyage that Columbus wrote this letter to Luis de Sant Angel, treasurer of Aragon, who had given him substantial help in fitting

out his expedition. This announcement of his discovery of the West Indies was evidently intended for the eyes of Ferdinand and Isabella. The text of the present translation can be read on the "Early Modern Spain" website maintained by the Departments of Spanish and Spanish-American Studies, King's College London, available at http://www.ems.kcl.ac.uk/content/etext/e022.html.

2 See my *Making Waves: The Continuing Portuguese Adventure*, 19.

3 William Davies Durland, "The Forests of the Dominican Republic," *Geographical Review* 12, no. 2 (1922), https://www.jstor.org/stable/208737?seq=1#metadata_info_tab_contents.

4 Suzanne Austin Alchon, *A Pest in the Land: New World Epidemics in a Global Perspective* (Albuquerque: University of New Mexico Press, 2003), 85–108.

5 For a short, oversimplified history, see the course outline prepared by Bob Corbett, professor at Webster University, available at http://faculty.webster.edu/corbetre/haiti/history/course/unitone/short.htm.

6 See "The First Colony," in Richard A. Haggerty, ed., *Dominican Republic: A Country Study* (Washington, DC: GPO for the Library of Congress, 1989), available at http://countrystudies.us/dominican-republic/3.htm.

7 "Haiti and Santo Domingo," in Haggerty, *Dominican Republic.* "Colonial Society," in Richard A. Haggerty, ed., *Haiti: A Country Study* (Washington, DC: GPO for the Library of Congress, 1989), http://countrystudies.us/haiti/8.htm.

8 John Lynch, *Simón Bolívar: A Life* (New Haven, CT: Yale University Press, 2006), 36.

9 Conversion tables are available at http://www.histoire-genealogie.com/spip.php?article3988&lang=fr.

10 Laurent Dubois, *Haiti: The Aftershocks of History* (New York: Picador, 2003), 8.

11 Haiti and Santo Domingo," in Haggerty, *Dominican Republic.*

12 For the treaty between the United States and the Dominican Republic, see "Dominican Republic-United States," *American Journal of International Law* 36, no. 4 (1942): 209–13.

13 Richard Lee Turits, "A World Destroyed, A Nation Imposed: The 1937 Haitian Massacre in the Dominican Republic," *Hispanic American Historical Review* 82, no. 3 (2002): 613.

14 This dark history has its biblical echoes:

> And the Gileadites took the passages of Jordan before the Ephraimites: and it was so, that when those Ephraimites which were escaped said, Let me go over; that the men of Gilead said unto him, Art thou an Ephraimite? If he said, Nay; Then said they unto him, Say now Shibboleth: and he said Sibboleth: for he could not frame to pronounce it right. Then they took him, and slew him at the passages of Jordan: and there fell at that time of the Ephraimites forty and two thousand.

From Judges 12:5.

15 "The legacy of Dr. José Francisco Peña Gómez," (press release), *Dominican Today*, April 19, 2009, available at https://web.archive.org/web/20090422000621

http://www.dominicantoday.com/dr/local/2009/4/19/31731/The-legacy-of-Dr-Jose-Francisco-Pena-Gomez.

16 Michele Wucker, "The Dominican Republic's Shameful Deportation Legacy," *Foreign Policy*, October 8, 2015, http://foreignpolicy.com/2015/10/08/dominican-republic-haiti-trujillo-immigration-deportation/.

17 Amnesty International, *"Where Are We Going to Live?" Migration and Statelessness in Haiti and the Dominican Republic* (London: Amnesty International, 2016), 4; the report can be viewed at https://www.amnesty.org/en/documents/amr36/4105/2016/en/.

18 See UN Office for the Coordination of Humanitarian Affairs, "Haiti: Humanitarian Snapshot, as of 31 August 2018," https://reliefweb.int/report/haiti/haiti-humanitarian-snapshot-31-august-2018.

19 Richard André, "The Dominican Republic and Haiti: A Shared View from the Diaspora. A conversation with Edwidge Danticat and Junot Díaz," *Americas Quarterly* (Summer 2014), http://americasquarterly.org/content/dominican-republic-and-haiti-shared-view-diaspora.

20 Edwidge Danticat, *The Farming of Bones* (New York: Soho Press, 1998), 266.

21 Mario Vargas Llosa, *The Feast of the Goat*, trans. Edith Grossman (New York: Picador, 2005), 143.

22 John W. Graham, *Whose Man in Havana? Adventures from the Far Side of Diplomacy* (Calgary: University of Calgary Press, 2015), 9.

23 Dubois, *Haiti*, 314–31 .

24 Sean Mills, *A Place in the Sun: Haiti, Haitians, and the Remaking of Quebec* (Montreal: McGill-Queen's University Press, 2016), 210.

25 See Albert Valdman, "Creole: The National Language of Haiti," *Footsteps* 2, no. 4 (2000), http://www.indiana.edu/~creole/creolenatllangofhaiti.html.

26 Oliver Wainwright, "In Iceland, 'respect the elves—or else'," *Guardian*, March 25, 2015, http://www.theguardian.com/artanddesign/2015/mar/25/iceland-construction-respect-elves-or-else.

27 See "One Island, Two Peoples, Two Histories: The Dominican Republic and Haiti," chapter 11 of Jared Diamond's *Collapse: How Societies Choose to Fail or Succeed* (New York: Viking Press, 2005).

28 Ibid., 344.

29 "Haiti: Jovenel Moise confirmed as new president," *Al Jazeera*, April 11, 2017, http://www.aljazeera.com/news/2017/01/haiti-jovenel-moise-president-170104054434935.html; Agence France-Presse, "Jovenel Moïse élu président d'Haïti" 29 novembre 2016," *Le Devoir*, November 29, 2016, http://www.ledevoir.com/international/actualites-internationales/485865/haiti-jovenel-moise-elu-president-d-haiti.

30 Daniel Zovatto, "Dominican Republic opts for continuity," *Brookings*, June 2, 2016, https://www.brookings.edu/opinions/dominican-republic-opts-for-continuity-2/.

31 See "Countries of Birth for U.S. Immigrants, 1960-Present," Migration Policy Institute, https://www.migrationpolicy.org/programs/data-hub/charts/immigrants-countries-birth-over-time?width=1000&height=850&iframe=true.

32 Statistics Canada, "Census Profile, 2016 Census," 2017, https://www12.statcan. gc.ca/census-recensement/2016/dp-pd/prof/index.cfm?Lang=E.

33 Mills, *A Place in the Sun*, 135.

34 Stephen Azzi, "Michaëlle Jean," *Canadian Encyclopedia*, November 15, 2010, http:// www.thecanadianencyclopedia.com/en/article/michaelle-jean/.

35 Wasiq N. Khan, "Economic Growth and Decline in Comparative Perspective: Haiti and the Dominican Republic, 1930–1986," *Journal of Haitian Studies* 16, no. 1 (2010): 112–25; see also "Dominican Republic vs. Haiti," *Index Mundi*, https://www. indexmundi.com/factbook/compare/dominican-republic.haiti.

36 Dubois, *Haiti*, 12.

37 Coinage N. Gothard, "Audit of USAID/Haiti's Renovation of of the Beau Rivage Hotel for Use as a Mission Compound" (Memorandum Report No.-1-521-86-02, November 22, 1985), available at https://pdf.usaid.gov/pdf_docs/PDAAS133.pdf.

38 The photos are available at http://www.nytimes.com/interactive/world/haiti-panoramas.html#/o.

39 See Canadian Press, "L'oncle de la ministre Anglade assassiné en Haïti," *Le Devoir*, April 11, 2016, http://www.ledevoir.com/international/actualites-internationales/ 467830/l-oncle-de-la-ministre-anglade-assassine-en-haiti.

7: BURUNDI AND RWANDA

1 My trip to Africa was sponsored by the Conseil des arts et lettres du Québec, which gave me a grant to research my novel *The Violets of Usambara* (Toronto: Cormorant Books, 2008). The book is about a Canadian politician who is part of a fact-finding team investigating refugee camps in Burundi in 1997. I took extensive notes, but when I pulled them out to work on this chapter, I wasn't quite sure how much I could count on their accuracy, or that of the vivid memories they conjured up. Then I realized I had a partial test in the transcript of the trial that I witnessed at the Arusha International Tribunal. To my great delight it corroborated in detail what I remembered. I've therefore decided to trust my memories here.

2 Figures from *World Population Review*, available at http://worldpopulationreview. com/countries/burundi-population/ and http://worldpopulationreview.com/ countries/rwanda-population/.

3 See "Netherlands agricultural exports hit new high," *Netherland Times*, January 26, 2016, http://www.nltimes.nl/2016/01/26/netherlands-agricultural-exports-hit-new-high/, and Government of Netherlands, "Agricultural exports worth nearly €92 billion in 2017," January 19, 2018, https://www.government.nl/latest/news/2018/01/ 19/agricultural-exports-worth-nearly-%E2%82%AC92-billion-in-2017.

4 Germain Bayon et al., "Intensifying Weathering and Land Use in Iron Age Central Africa," *Science* 335, no. 6073 (2012), http://dx.doi.org/10.1126/science.1215400.

5   John Thornton, *Africa and Africans in the Making of the Atlantic World, 1400–1800* (Cambridge: Cambridge University Press, 1998), 46.

6   Bayon, "Intensifying."

7   Michela Leonardi et al., "The evolution of lactase persistence in Europe. A synthesis of archaeological and genetic evidence," *International Dairy Journal* 22, no. 2 (2012), https://dx.doi.org/10.1016%2Fj.idairyj.2011.10.010.

8   Sarah A Tishkoff et al., "Convergent adaptation of human lactase persistence in Africa and Europe," *Nature Genetics* 39, no. 1 (2007), https://doi.org/10.1038/ng1946.

9   Jared Diamond, *Guns, Germs, and Steel: The Fates of Human Societies* (New York and London: W.W. Norton & Company, 1999), 328.

10 There are two notable exceptions: Guarani in Paraguay and Quechua in Peru and Bolivia, which are both official or co-official languages in their respective countries. As it happens, the latter is a language of conquest, since it was imposed by the Inca as they built their Andean empire.

11 John Hanning Speke, *Journal of the Discovery of the Source of the Nile* (London: W. Blackwood and Sons, 1863). The passage quoted here comes from chapter 9, "History of the Wahuma," available from Project Gutenberg at http://www.gutenberg.org/files/3284/3284.txt.

12 Beniamino Trombetta et al., "Phylogeographic refinement and large scale genotyping of human Y chromosome haplogroup E provide new insights into the dispersal of early pastoralists in the African continent," *Genome Biology and Evolution* 7, no. 7 (2015), https://doi.org/10.1093/gbe/evv118.

13 "In the East of Africa," *New York Times*, February 13, 1964, https://www.nytimes.com/1964/02/13/archives/in-the-east-of-africa.html.

14 René Lemarchand, *Burundi: Ethnic Conflict and Genocide* (Cambridge and New York: Woodrow Wilson Center Press and Cambridge University Press, 1996), 16.

15 Lemarchand, *Burundi*, 32.

16 Jean-Pierre Chrétien, *L'Afrique des Grands Lacs: Deux mille ans d'histoire* (Paris: Aubier, 2000), 38.

17 Ibid., 263.

18 Ibid., 167.

19 Lemarchand, *Burundi*, 51.

20 Ibid., 51.

21 Ibid., 52.

22 In June and July of 1962, the *New York Times* published twenty-five news stories about Rwanda's and Burundi's independence, and thereafter regularly reported on events in the two countries.

23 For a good profile of Paul Kagame, see Richard Grant, "Paul Kagame: Rwanda's redeemer or ruthless dictator?" *Telegraph*, July 22, 2010, https://www.telegraph.co.uk/news/worldnews/africaandindianocean/rwanda/7900680/Paul-Kagame-Rwandas-redeemer-or-ruthless-dictator.html.

24 Because I did not flag my exchanges with members of this family as information that I might use in a future non-fiction work, I'm not using their names here so as to protect their privacy.

25 Tor Sellstrom and Lennart Wohlgemuth, *The International Response to Conflict and Genocide: Lessons from the Rwanda Experience* (JEEAR) (Uppsala, SE: The Nordic Africa Institute, 1996).

26 Some sources say that Kagame was actually the instigator of the attack on the airplane. In a case brought in France by the families of the plane's civilian crew, one of Kagame's close associates charged that he admitted as much on the day of the attack. But after years of investigation and testimony, in October 2018 a French prosecutor recommended dropping charges against eight senior Rwandan officials because insufficient evidence had been presented to warrant further legal action. See Agence France-Presse, "Rwanda: non-lieu requis en France dans l'enquête sur l'attentat déclencheur du génocide," *L'Express*, October 13, 2018, https://www.lexpress.fr/actualites/1/societe/rwanda-non-lieu-requis-en-france-dans-l-enquete-sur-l-attentat-declencheur-du-genocide_2039706.html.

27 The commission's final report, released in 2004, is available at https://www.usip.org/sites/default/files/file/resources/collections/commissions/Burundi-Report.pdf.

28 See Pacifique Cubahiro, "La première session du Tribunal Russell sur le Burundi sera organisée à Paris en septembre," *Infos Grands Lacs*, August 27, 2015, http://infosgrandslacs.info/productions/la-premiere-session-du-tribunal-russell-sur-le-burundi-sera-organisee-paris-en-septembre.

29 This comes from "Background Information on the Justice and Reconciliation Process in Rwanda," available from the Outreach Programme on the Rwanda Genocide and the United Nations at http://www.un.org/en/preventgenocide/rwanda/about/bgjustice.shtml.

30 For a more complete view of the reconciliation process, see "Rwanda Reconciliation," an audio documentary first broadcast on CBC Radio on December 8, 2014, and available at http://www.cbc.ca/radio/ideas/rwanda-reconciliation-1.2914211.

31 I've always taken extensive notes on my travels, and been pretty confident that what I recorded was as near the truth of what was transpiring as humanly possible. But you never know when memory and/or faulty note-taking might play tricks on you. That's why I was delighted to discover that testimony from this trial is now available online by request. I duly received a transcript of what I'd heard on October 19, 2001 (classified as International Criminal Tribunal for Rwanda Case No. ICTR-96-10-T ICTR-96-17-T). What I found there confirmed both my notes and my memories.

32 Grant, "Paul Kagame."

33 Amnesty International, "Rwanda: Decades of attacks repression and killings set the scene for next month's election," July 7, 2017, https://www.amnesty.org/en/

latest/news/2017/07/rwanda-decades-of-attacks-repression-and-killings-set-the-scene-for-next-months-election/.

34 Will Ferguson, *Road Trip Rwanda: A Journey into the New Heart of Africa* (Toronto: Viking, 2005), 50.

35 Gaël Faye, *Petit Pays* (Paris: French and European Publications, 2017), 170. My translation.

36 Edmund Kagire, "Rwanda's Mushikiwabo takes the reins at Francophone club of 58 countries," *East African*, October 13, 2018, http://www.theeastafrican.co.ke/news/ea/Rwanda-Mushikiwabo-takes-the-reins-at-Francophonie/4552908-4803792-12l17j6z/index.html.

37 See "Corruption Perceptions Index 2017," available at https://www.transparency.org/news/feature/corruption_perceptions_index_2017, and "Burundi GDP per capita," *Trading Economics*, 2019, https://tradingeconomics.com/burundi/gdp-per-capita.

38 "Burundi: pourquoi Pierre Nkurunziza a-t-il annoncé son départ?" *Radio France International*, June 9, 2018, http://www.rfi.fr/afrique/20180608-burundi-nkurunziza-vraiment-partir.

39 International Federation for Human Rights, *Burundi on the brink, looking back on two years of terror* (Report No. 693a, June 2017), https://www.fidh.org/IMG/pdf/burundi_jointreport_june2017_eng_final.pdf.

8: SCOTLAND AND IRELAND

1 I must admit that I'm not sure of his first name after all these years, but he definitely was an O'Neill.

2 Mihai Andrei, "Doggerland—the land that connected Europe and the UK 8000 years ago," *ZME Science*, February 16, 2017, http://www.zmescience.com/science/geology/doggerland-europe-land/.

3 Megan Lane, "The moment Britain became an island," *BBC News Magazine*, February 15, 2011, https://www.bbc.com/news/magazine-12244964.

4 Some helpful background on Newgrange is available at http://www.newgrange.com/michael-j-okelly.htm.

5 Roff Smith, "Before Stonehenge," *National Geographic*, August 2014, http://ngm.nationalgeographic.com/2014/08/neolithic-orkney/smith-text.

6 Ibid.

7 Lara M. Cassidy et al., "Neolithic and Bronze Age migration to Ireland and establishment of the insular Atlantic genome," *PNAS* 113, no. 2 (2016), https://doi.org/10.1073/pnas.1518445113.

8 Jeanna Bryner, "One Common Ancestor Behind Blue Eyes," *Live Science*, January 31, 2008, http://www.livescience.com/9578-common-ancestor-blue-eyes.html, and Hans Eiberg et al., "Blue eye color in humans may be caused by a perfectly

NOTES | 227

associated founder mutation in a regulatory element located within the HERC2 gene inhibiting OCA2 expression," *Human Genetics* 123, no. 2 (2008), https://doi.org/10.1007/s00439-007-0460-x.

9   Paul Rincon, "Ancient Britons 'replaced' by newcomers," BBC *News*, February 21, 2018, https://www.bbc.com/news/science-environment-43115485, and Iñigo Olalde et al., "The Beaker phenomenon and the genomic transformation of northwest Europe," *Nature* 555 (2018), https://doi.org/10.1038/nature25738.

10 Tim Radford, "Irish DNA originated in Middle East and eastern Europe," *Guardian*, December 28, 2015, https://www.theguardian.com/science/2015/dec/28/origins-of-the-irish-down-to-mass-migration-ancient-dna-confirms, and Cassidy et al., "Neolithic and Bronze Age migration."

11 Mallick et al., "The Light Skin Allele of SLC24A5."

12 Called MC1R (melanocortin 1 receptor), the gene is one of several governing production of pigment in skin, eyes, and hair. It exists in many variations: the ones that produce red hair effectively stop the production of brown pigment—eumelanin—and allow phenomelanin, which is red-yellow, to dominate. See Gregory S. Barsh, "What Controls Variation in Human Skin Color?" PLOS *Biology* 1, no. 1 (2003), https://doi.org/10.1371/journal.pbio.0000027.

13 See "BritainsDNA Announces the Results of the Red-Head Project," available at https://studylib.net/doc/13867454/britainsdna-announces-the-results-of-the-red-head-project.

14 Herodotus, *The Histories*, tr. Aubrey de Sélincourt (London: Penguin, 1972), bk. 4, ch. 108: 276.

15 Chunxiang Li et al., "Analysis of ancient human mitochondrial DNA from the Xiaohe cemetery: Insights into prehistoric population movements in the Tarim Basin, China," BMC *Genetics* 16, no. 78 (2015), https://doi.org/10.1186/s12863-015-0237-5.

16 See Elizabeth Wayland Barber, *The Mummies of Ürümchi* (New York: W.W. Norton & Company, 2004), 144–5.

17 Julius Caesar, *The Gallic Wars*, trans. W.A. McDevitte and W.S. Bohn, bk. 1, ch. 29, and bk. 6, ch. 14, available at http://classics.mit.edu/Caesar/gallic.html.

18 For a good overview of the Celts and their civilization, see the three-part BBC documentary *The Celts: Blood, Iron and Sacrifice*, available at https://www.youtube.com/watch?v=zA-itb5NwDU&t=149s awzXYvvQ7MU. See also the citation making Hallstatt-Dachstein a UNESCO World Heritage Site, at http://whc.unesco.org/en/list/806. For more about the La Tène site in Switzerland, see the document outlining museum exhibitions on the 150th anniversary of the discovery of the remains of Celtic culture there, available at http://www.swissinfo.ch/eng/unearthing-la-t%C3%A8ne-s-celtic-mysteries/5776464.

19 Quote from Earl A. Powell III, director of the National Gallery of Art in Washington, DC, where the statue was on display in 2014. See "National Gallery of Art, Roma Capitale, and the Embassy of Italy in Washington, DC, Present *The*

Dying Gaul: An Ancient Roman Masterpiece from the Capitoline Museum, Rome," November 26, 2013, http://www.nga.gov/content/ngaweb/press/exh/3655.html.

20 Ceasar, *Gallic Wars*, ch. 16.

21 See Strabo, *The Geography*, bk. 4, ch. 4 (c. 7 BCE), available at http://penelope. uchicago.edu/Thayer/E/Roman/Texts/Strabo/4D*.html. Strabo lived from 64 BCE to 24 CE.

22 Thomas Cahill, *How the Irish Saved Civilization: The Untold Story of Ireland's Heroic Role from the Fall of Rome to the Rise of Medieval Europe* (New York: Doubleday, 1995), 123.

23 Silvia M. Bello et al., "An Upper Palaeolithic engraved human bone associated with ritualistic cannibalism," PLOS ONE 12, no. 8 (2017), https://doi.org/10.1371/journal.pone.0182127.

24 *The Confession of St. Patrick* gives his version of what happened to him, and why he returned to Ireland to preach and convert. It is available in a Project Gutenberg edition of *The Most Ancient Lives of Saint Patrick* (ed. James O'Leary), at http://www.gutenberg.org/files/18482/18482-h/18482-h.htm#chap01.

25 Cahill, *How the Irish Saved Civilization*, 27.

26 From Ken McGoogan, *Celtic Lightning: How the Scots and the Irish Created a Canadian Nation* (Toronto: Patrick Crean Editions, 2015), 82.

27 Ibid., 88. And I should mention here that as a university student I had the pleasure of being compared to Queen Boudicea, the Celtic heroine, by a professor: "She must have looked like you, Madame, tall, strong, fair and red-haired."

28 Cahill, *How the Irish Saved Civilization*, 170–1.

29 For an informative digest of the history of the British Isles and Ireland, see the handsome volume by R.G. Grant et al., *History of Britain and Ireland: The Definitive Visual Guide* (New York: DK Publishers, 2011).

30 Ben Kiernan, *Blood and Soil: A World History of Genocide and Extermination from Sparta to Darfur* (New Haven, CT: Yale University Press, 2007), 169–212.

31 Ibid., 179.

32 Ibid., 182.

33 See Matthew Engel, "Northern Ireland: An uncertain peace," *Guardian*, January 21, 2017, https://www.theguardian.com/uk-news/2017/jan/21/northern-ireland-an-uncertain-peace.

34 For some helpful background, see "Good Friday Agreement," *Encyclopaedia Britannica* (2019), https://www.britannica.com/topic/Good-Friday-Agreement.

35 "Shankill bonfire: Belfast terraced houses destroyed by blaze," BBC News, July 12, 2016, http://www.bbc.com/news/uk-northern-ireland-36771384.

36 "The Twelfth: Thousands march in Orange Order parades," BBC News, July 12, 2016, https://www.bbc.com/news/uk-northern-ireland-36765294.

37 Martin Luther's Ninety-Five Theses are available at http://www.luther.de/en/95thesen.html. Today scholars doubt whether he actually took a hammer and did the nailing. The story appears only after his death, but it is certain

that on October 31, 1517, he wrote a letter to his superiors denouncing religious corruption.

38 For more on Catherine's story, see "Catherine of Aragon: Queen of England," *Encyclopaedia Britannica* (2019), https://www.britannica.com/biography/Catherine-of-Aragon.

39 Some historical background on the Church of Scotland is available at http://www.churchofscotland.org.uk/about_us/how_we_are_organised/history.

40 Arthur Herman, *How the Scots Invented the Modern World: The True Story of How Western Europe's Poorest Nation Created Our World and Everything in It* (New York: Broadway Books, 2002), 22.

41 John Scally, "Why did the Reformation fail to take hold in an Ireland under English rule?" *Irish Times*, October 10, 2017, https://www.irishtimes.com/opinion/why-did-the-reformation-fail-to-take-hold-in-an-ireland-under-english-rule-1.3249598.

42 Herman, *How the Scots Invented the Modern World*, 11.

43 See Wesley Johnston, "Prelude to Famine," available at http://www.wesleyjohnston.com/users/ireland/past/famine/demographics_pre.html.

44 Herman, *How the Scots Invented the Modern World*, 231.

45 The Highland Clearances was the eviction of a large number of tenant farmers in the Scottish Highlands beginning in the late eighteenth century. Common lands were enclosed and the economy shifted from subsistence agriculture to sheep raising. For some background, see Ross Noble, "The Cultural Impact of the Highland Clearances," BBC, February 17, 2011, http://www.bbc.co.uk/history/british/civil_war_revolution/scotland_clearances_01.shtml.

46 Herman, *How the Scots Invented the Modern World*, 338.

47 See Angus Maddison, "The Contours of World Development," available at http://www.ggdc.net/maddison/other_books/Ch.1_2001.pdf. This series of data covering more than a thousand years is an impressive source of information about the economic state of the world and its population.

48 This from the 2005 Gaelic Language (Scotland) Act, the text of which is available at http://www.legislation.gov.uk/asp/2005/7/contents#1.

49 Inayat Shah, "Linguistic Attitude and the Failure of Irish Language Revival Efforts," *International Journal of Innovation and Scientific Research* 1, no. 2 (2014), http://www.ijisr.issr-journals.org/abstract.php?article=IJISR-14-111-02.

50 Figures available at https://faithsurvey.co.uk/irish-census.html.

51 See Joel Gunter, "Abortion in Ireland: The fight for choice," *BBC News*, March 8, 2017, http://www.bbc.com/news/world-europe-39183423.

52 Matt Schiavenza, " 'Well Done': The Legalization of Gay Marriage in Ireland," *Atlantic*, May 23, 2015, https://www.theatlantic.com/international/archive/2015/05/ireland-gay-marriage/394052/.

53 Henry McDonald, "Ireland's first gay prime minister Leo Varadkar formally elected," *Guardian*, June 14, 2017, https://www.theguardian.com/world/2017/jun/14/leo-varadkar-formally-elected-as-prime-minister-of-ireland.

54 "Irish abortion referendum: Ireland overturns abortion ban," BBC News, May 26, 2018, https://www.bbc.com/news/world-europe-44256152.

55 Judith Duffy, "Welcome to Secular Scotland...a nation where religion is in retreat," Herald, May 29, 2016, http://www.heraldscotland.com/news/14523231. Welcome_to_Secular_Scotland____a_nation_where_religion_is_in_retreat/; see also, Government of Scotland, "Summary: Religion Demographics," 2018, http://www.gov.scot/Topics/People/Equality/Equalities/DataGrid/Religion/RelPopMig.

56 See "Ireland—Unemployment rate," Knoema, https://knoema.com/atlas/Ireland/Unemployment-rate.

57 John FitzGerald, "In Ireland, free education has more than proved its worth," Irish Times, March 24, 2017, https://www.irishtimes.com/business/economy/in-ireland-free-education-has-more-than-proved-its-worth-1.3022196.

58 See "The Celtic Tiger," http://www.mtholyoke.edu/~falve22h/classweb/recession/recession/Celtic_Tiger.html.

59 "Ireland Unemployment Rate," Trading Economics, 2019, https://tradingeconomics.com/ireland/unemployment-rate.

60 "Scottish unemployment rate hits record low at 3.7%," BBC News, December 11, 2018, https://www.bbc.com/news/uk-scotland-46522798.

61 "Scottish Government writes to MPS ahead of crucial Brexit vote," South Wales Guardian, January 12, 2019, https://www.southwalesguardian.co.uk/news/national/17354263.scottish-government-writes-to-mps-ahead-of-crucial-brexit-vote/.

62 Euan McKirdy, "Scottish referendum in doubt after steep losses for SNP in UK vote," CNN, June 9, 2017, http://www.cnn.com/2017/06/08/europe/snp-uk-general-election/index.html.

63 George Parker, Jim Pickard, and Arthur Beesley, "Theresa May wins Queen's Speech vote with slender majority," Financial Times, June 29, 2017, https://www.ft.com/content/6ddfc7a0-5cd7-11e7-b553-e2df1b0c3220?mhq5j=e3. See also, Jonathan S. Blake, "What a Protestant Parade Reveals About Theresa May's New Partners," Atlantic, July 11, 2017, https://www.theatlantic.com/international/archive/2017/07/protestant-parade-northern-ireland/533151/.

64 Joe Duggan, "Confidence vote latest: DUP 'DELIGHTED' to hold balance of power—May RELIANT on NI party," Express, January 16, 2019, https://www.express.co.uk/news/politics/1073315/confidence-vote-brexit-news-DUP-theresa-may-prime-minister.

9: VERMONT AND NEW HAMPSHIRE

1 See United States Census Bureau, "QuickFacts: Maine; New Hampshire; Vermont; United States," at https://www.census.gov/quickfacts/fact/table/me,nh,vt,US/PST045217, and "U.S. States comparison: Vermont vs New Hampshire," at https://countryeconomy.com/countries/usa-states/compare/vermont/new-hampshire.

2   Calculated from US Census Bureau statistics; see https://www.census.gov/
    quickfacts/vt and https://www.census.gov/quickfacts/nh.
3   United States Census Bureau, "2010 Census Shows Second Highest
    Homeownership Rate on Record Despite Largest Decrease since 1940"
    (press release), October 6, 2011, https://www.census.gov/newsroom/releases/
    archives/2010_census/cb11-cn188.html.
4   See "New Hampshire," *270 to Win*, 2019, http://www.270towin.com/states/New_
    Hampshire.
5   For election results, see "New Hampshire Results," *New York Times*, August 1,
    2017, http://www.nytimes.com/elections/results/new-hampshire.
6   For a fascinating account of Vermont's geological history, which also touches
    on that of New Hampshire, see Nancy Bazilchuk and Rick Strimbeck, *Longstreet
    Highroad Guide to the Vermont Mountains* (Atlanta: Longstreet Press, 1997), which
    is available at https://www.nasw.org/users/nbazilchuk/Articles/ch1web.htm.
7   Ibid.
8   Frederick W. Kilbourne, *Chronicles of the White Mountains* (Boston and New York:
    Houghton Mifflin, 1916), 63.
9   The earliest evidence of what some anthropologists call Paleoindians in the
    region date from about 8,000 BCE. See Jonathan C. Lothrop et al., "Paleoindians
    and the Younger Dryas in the New England-Maritimes Region," *Quaternary
    International* 242, no. 2 (2011), https://doi.org/10.1016/j.quaint.2011.04.015.
10 See Alexander Pfaff, "From Deforestation to Reforestation in New England,
    United States," in *World Forests from Deforestation to Transition? World Forests*, vol.
    2, ed. M. Palo and H. Vanhanen (Dordrecht, NL: Springer, 2000).
11 The Massachusetts Charter was proclaimed in 1691, and applied to New
    Hampshire, which had been made a separate colony in 1679.
12 Confusion between "Abenaki" and "Wabanki," the name of a larger grouping of
    several societies living from the Canadian Maritimes through the Champlain
    Valley, has blurred the distinction. This group had patrilineal society, unlike their
    neighbours (and frequent enemies) the Iroquois, who were matrilineal with
    inheritance through the mothers' lines.
13 See "Abenaki," *New World Encyclopedia*, 2016, http://www.newworldencyclopedia.
    org/entry/Abenaki.
14 James H. Marsh, "The 1704 Raid on Deerfield," *Canadian Encyclopedia*, March
    4, 2015, http://www.thecanadianencyclopedia.ca/en/article/the-deerfield-raid-
    feature/; see also Caroline Monpetit, "Une captive chez les Iroquois," *Le Devoir*,
    March 11, 2005, http://www.ledevoir.com/societe/actualites-en-societe/76706/
    une-captive-chez-les-iroquois.
15 Mark A. Peterson, *The Price of Redemption: The Spiritual Economy of Puritan New
    England* (Palo Alto, CA: Stanford University Press, 1997), 232.
16 For an engaging look at the Allen family's move from Connecticut to—
    eventually—the land near what is now Burlington, Vermont, see Michael A.

Bellesiles's *Revolutionary Outlaws: Ethan Allen and the Struggle for Independence on the Early American Frontier* (Charlottesville: University Press of Virginia, 1993).

17 Ibid., 23.

18 Ibid., 140.

19 For Twilight's biography, see Old Stone House Museum, "Alexander Twilight," available at http://oldstonehousemuseum.org/twilight-bio/.

20 Jack Zeilenga, "Women in VT Politics: During and Post Suffrage 1840–1940," *Vermont Women's History Project*, https://vermonthistory.org/images/stories/vwhp/Women%20in%20Politics.pdf.

21 See J. Dennis Robinson, "Whittier's Anti-Slavery Ode to NH," available at http://www.seacoastnh.com/whittiers-anti-slavery-ode-to-nh/.

22 Chuck Wooster, "Vermont—New Hampshire: The Hydropower Difference," *Northern Woodlands*, April 14, 2002, https://northernwoodlands.org/outside_story/article/vermont-new-hampshire-the-hydropower-difference.

23 See Bazilchuk and Strimbeck, *The Vermont Mountains*, ch. 1.

24 Chuck Wooster, "Glaciers and Taxes in Vermont and New Hampshire," *Northern Woodlands*, May 12, 2002, https://northernwoodlands.org/outside_story/article/glaciers-and-taxes-in-vermont-and-new-hampshire.

25 See Dan Caplinger, "The 5 States With No Sales Tax," AOL.*com*, October 17, 2014, http://www.aol.com/article/2013/05/05/the-5-states-with-no-sales-tax/20558414/.

26 Wooster, "Glaciers and Taxes."

27 Kenneth Roberts, *Northwest Passage* (Garden City, NY: Doubleday, Doran and Co., 1937), 28.

28 Quoted in Corydon Ireland, "Vermont and New Hampshire, geographic twins, cultural aliens," *Harvard Gazette*, November 1, 2007, http://news.harvard.edu/gazette/story/2007/11/vermont-and-new-hampshire-geographic-twins-cultural-aliens/.

29 See, for example, http://www.city-data.com/forum/new-hampshire/1181182-looking-honest-comparison-vermont-vs-new.html.

30 Arnie Arnesen's February 10, 2016 interview with Democracy Now is available at https://www.youtube.com/watch?v=tsMF_7XVPSU.

31 For some background, see Woodsmoke Productions and Vermont Historical Society, "The Flood of '27," *The Green Mountain Chronicles* radio broadcast and background information, original broadcast 1988-89, accessed on the web at http://vermonthistory.org/research/research-resources-online/green-mountain-chronicles/the-flood-of-27-1927.

32 A fascinating analysis can be found in Blake Harrison's *The View from Vermont: Tourism and the Making of an American Rural Landscape* (Burlington: University of Vermont Press, 2006).

33 Kilbourne, *Chronicles*, ix.

34 Much of the following description is taken from an article I wrote about the visit: "Old Time Fiddlers Make Beautiful Music in Vermont," *New York Times*, May 6, 1973.

35 The Liberty Union Party went out of business in 1976.

36 Paul Heintz, "Bernie Sanders Declines Democratic Senatorial Nomination," *Seven Days*, August 21, 2018, https://www.sevendaysvt.com/OffMessage/ archives/2018/08/21/bernie-sanders-declines-democratic-senatorial-nomination.

37 Figures from "Vermont Election Results 2018," *Politico*, March 14, 2019, https:// www.politico.com/election-results/2018/vermont/.

38 See April McCullum, "Who won the Vermont governor election and what that means," *Burlington Free Press*, November 7, 2018, https://www.burlingtonfreepress. com/story/news/politics/elections/2018/11/07/vermont-midterm-election-results-what-governors-race-means-phil-scott-christine-hallquist/1806899002/.

39 Todd Bookman, "Democrats Retake Both Chambers of New Hampshire Legislature," *New Hampshire Public Radio*, November 7, 2018, http://www.nhpr.org/ post/democrats-retake-both-chambers-new-hampshire-legislature#stream/0.

40 See "New Hampshire Opioid Summary," National Institute on Drug Abuse, February 2018, https://www.drugabuse.gov/drugs-abuse/opioids/opioid-summaries-by-state/new-hampshire-opioid-summary. See also Casey Leins, "New Hampshire: Ground Zero for Opioids," *US News & World Report*, June 28, 2017, https://www.usnews.com/news/best-states/articles/2017-06-28/ why-new-hampshire-has-one-of-the-highest-rates-of-opioid-related-deaths, and German Lopez, "I looked for a state that's taken the opioid epidemic seriously. I found Vermont," *Vox*, October 31, 2017, https://www.vox.com/ policy-and-politics/2017/10/30/16339672/opioid-epidemic-vermont-hub-spoke.

10: ALBERTA AND SASKATCHEWAN

1 Wallace Stegner, *Wolf Willow: A History, a Story and a Memory of the Last Plains Frontier* (New York: Viking Press, 1962), 6–7.

2 Basically, some members of the political establishment in Central Canada didn't want to create one province because it might carry too much political weight.

3 The speech was recorded by the CBC, but never broadcast. I've tried to find it in the archives, but have had to rely on my notes from the evening, and on clips from speeches that Douglas gave during this period. An example can be found at https://www.youtube.com/watch?v=C2oUInTUIAM.

4 Walter Stewart, *The Life and Political Times of Tommy Douglas* (Toronto: McArthur and Company, 2003), 50.

5 In the second stage of the CBC program, ten men (and significantly no women) were in the competition. They included two great inventors, Alexander Graham Bell and Dr. Frederick Banting (a co-discoverer of insulin), a couple of prime ministers, but also a hockey player and a sports commentator. See "And the Greatest Canadian of all time is…" *CBC Television*, November 29, 2004, available at http://www.cbc.ca/archives/entry/and-the-greatest-canadian-of-all-time-is.

6   See "History of Canada's Public Health Care," *Canadian Health Coalition*, 2016, https://www.healthcoalition.ca/tools-and-resources/history-of-canadas-public-health-care/, and Lorne Brown and Doug Taylor, "The Birth of Medicare: From Saskatchewan's breakthrough to Canada-wide coverage," *Canadian Dimension* 46, no. 4 (2012), https://canadiandimension.com/articles/view/the-birth-of-medicare.

7   John Ibbitson, *Stephen Harper* (Toronto: Signal, 2015), 29–30. This excellent biography is the source of the following paragraphs. For Harper's view on the National Energy Program and his movement to the right, see particularly chapters 2 and 3 (pp. 21–59).

8   "Stephen Harper's most controversial quotes compiled—by Tories," *Toronto Star*, April 25, 2011, https://www.thestar.com/news/canada/2011/04/25/stephen_harpers_most_controversial_quotes_compiled_by_tories.html.

9   Bill Waiser, *A World We Have Lost: Saskatchewan Before 1905* (Markham, ON: Fifth House Limited, 2016). This is a beautifully illustrated companion volume to Waiser's *Saskatchewan: A New History* (Calgary: Fifth House, 2005), which was published to mark the hundredth anniversary of Saskatchewan as a province.

10 "Hudson's Bay Company Beginnings," CBC *Learning*, 2001, http://www.cbc.ca/history/EPCONTENTSE1EP6CH1PA5LE.html.

11 Waiser, *A World We Have Lost*, 111–22.

12 Ibid., 183.

13 Ibid., 284.

14 J.M. Bumsted, "Red River Colony," *Canadian Encyclopedia*, March 25, 2015, http://www.thecanadianencyclopedia.ca/en/article/red-river-colony/.

15 Harper certainly took seriously the HBC claims, since they underlie Canada's recent claims to parts of the Arctic. The ill-fated Franklin expedition of 1845–8 aimed to find a northwest passage by going around islands that were collateral to the great land grant. Recent Canadian expeditions to search for the remains of Franklin's ships are "a stealthy way to exercise Canada's sovereignty over Arctic waters and to protect them from American, Russian and Chinese interests in shipping and energy development," according to Kat Long in "Stephen Harper's Franklin fever," *National Post*, May 22, 2014, http://nationalpost.com/opinion/kat-long-stephen-harpers-franklin-fever/wcm/f68dc0b2-e595-4745-a09c-dc7d5d235a04.

16 "Canada Buys Rupert's Land," CBC *Learning*, 2001, http://www.cbc.ca/history/EPCONTENTSE1EP9CH1PA3LE.html.

17 From James H. Marsh, "Railway History," *Canadian Encyclopedia*, March 4, 2015, http://www.thecanadianencyclopedia.ca/en/article/railway-history/.

18 Derek Hayes, "Drawing the lines," *Canadian Geographic* 125, no. 1 (2005): 48–9.

19 Gerald D. Nash, "The Census of 1890 and the Closing of the Frontier," *Pacific Northwest Quarterly* 71, no. 3 (1980): 98–100.

20 R. Douglas Francis and Tim B. Rogers, "Images of the Canadian West in the Settlement Era as Expressed in Song Texts of the Time," *Prairie Forum* 18, no. 2 (1993): 257–67.

21 In addition, German is pale yellow, light orange is Austrian, light green is
    Russian, red is Belgian, red and black is Dutch, light blue is "Hebrew," and
    orange is "Negro." The map in question can be viewed at http://ftp.geogratis.
    gc.ca/pub/nrcan_rncan/raster/atlas_2_ed/eng/peopleandsociety/population/
    page25_26.jpg.

22 James M. Pitsula, "Disparate Duo," *Beaver*, August/September (2005), https://
    canadashistory.partica.online/canadas-history/the-beaver-aug-sep-2005/
    flipbook/16/.

23 Stegner, *Wolf Willow*, 4.

24 André Munro, "Populism," *Encyclopaedia Britannica*, March 6, 2018, https://www.
    britannica.com/topic/populism.

25 Aidan Conway and J.F. Conway, "Saskatchewan: From Cradle of Social Democracy
    to Neoliberalism's Sandbox," in *Transforming Provincial Politics: The Political
    Economy of Canada's Provinces and Territories in the Neoliberal Era* (Studies in
    Comparative Political Economy and Public Policy), ed. Bryan M. Evans and
    Charles W. Smith (Toronto: University of Toronto Press, 2015), 226–54.

26 Ibid.

27 Seymour Martin Lipset, *Agrarian Socialism: The Cooperative Commonwealth
    Federation in Saskatchewan* (Berkeley: University of California Press, 1971), xix.

28 Ibid., 338.

29 Conway and Conway, "Saskatchewan," 114.

30 Ibid., 129.

31 David McGrane, "Explaining the Saskatchewan NDP's Shift to Third Way Social
    Democracy" (paper presented to the Canadian Political Science Association,
    Toronto, ON, June 3, 2006), https://www.cpsa-acsp.ca/papers-2006/McGrane-
    Explaining.pdf.

32 Mitchell Blair, "Elections Saskatchewan wants to change date of either 2020
    provincial or municipal election," *620 CKRM The Source*, May 31, 2017, http://www.
    620ckrm.com/2017/05/31/elections-saskatchewan-wants-to-change-date-of-either-
    2020-provincial-or-municipal-election/.

33 C.B. Macpherson, *Democracy in Alberta: Social Credit and the Party System*
    (Toronto: University of Toronto Press, 1953).

34 Danielle Smith, "Alberta already was an NDP province," *Globe and Mail*, May 8,
    2015, http://www.theglobeandmail.com/opinion/alberta-already-was-an-ndp-
    province/article24316040/.

35 Colby Cosh, "The death of the Alberta PC dynasty," *Maclean's*, May 7, 2015, https://
    www.macleans.ca/politics/the-death-of-the-alberta-pc-dynasty/.

36 Figures from Statistics Canada, available at https://www12.statcan.gc.ca/census-
    recensement/2016/as-sa/fogs-spg/Facts-PR-Eng.cfm?TOPIC=7&LANG=Eng&GK=
    PR&GC=47.

11: CANADA AND THE UNITED STATES OF AMERICA

1   Robyn Meredith, "5 Days in 1967 Still Shake Detroit," *New York Times*, July 23, 1997, http://www.nytimes.com/1997/07/23/us/5-days-in-1967-still-shake-detroit.html?pagewanted=all.

2   Footage of the Detroit riots is available at https://www.youtube.com/watch?v=jevooU3K9K8.

3   I am aware that this statement can—and probably will be—contested. But we're talking about shades of evil here, not the difference between the complete absence of racism and its polar opposite.

4   I should mention here again that my husband was not a draft dodger. He had deferments of various sorts from the time the war in Vietnam heated up until the draft lottery in 1969, when his number (236) gave him a very low priority. For some context, see "1969 U.S. Draft Lottery results," available at https://en.wikisource.org/wiki/User:Itai/1969_U.S._Draft_Lottery_results.

5   Lane Anker, "Peacekeeping and Public Opinion," *Canadian Military Journal*, June 20, 2005, http://www.journal.forces.gc.ca/vo6/no2/public-eng.asp.

6   Figures available at https://www.indexmundi.com/factbook/compare/united-states.canada/geography, https://www.worldometers.info/world-population/canada-population/, and http://www.census.gov/popclock/.

7   See OECD, "A Broken Social Elevator? How to Promote Social Mobility," *Report of the Organization for Economic Cooperation and Development*, 2018, https://www.oecd.org/social/soc/Social-mobility-2018-Overview-MainFindings.pdf, and "Canada," OECD *Better Life Index*, 2018, http://www.oecdbetterlifeindex.org/countries/canada/.

8   Audrea Lim, "Want to Move to Canada If Trump Wins? Not So Fast," *Rolling Stone*, May 5, 2016, http://www.rollingstone.com/politics/news/want-to-move-to-canada-if-trump-wins-not-so-fast-20160505.

9   "Robin Williams' Quote About Canada Will Make You Miss Him Even More," *Huffington Post Canada*, August 13, 2014, http://www.huffingtonpost.ca/2014/08/13/robin-williams-canada-quote-meth-lab_n_5676523.html.

10 Trudeau was speaking to the Washington Press Club in March 1969. See "The Elephant and the Mouse," *Parli: The Dictionary of Canadian Politics*, 2019, http://www.parli.ca/the-elephant-and-the-mouse/.

11 "The Château Ramezay during the American Invasion of 1775–1776," Château Ramezay—Historic Site and Museum of Montréal, https://www.chateauramezay.qc.ca/en/museum/history/american-revolution/.

12 In an attempt to rally the inhabitants of Quebec to the American cause, propaganda letters were translated into French and distributed in both Montreal and Quebec City in 1774, 1775, and 1776. Copies of these letters are available at http://www.axl.cefan.ulaval.ca/francophonie/HISTfrQC_S2_Brit-lettresUSA.htm.

13   Eli Meixler, " 'Didn't You Guys Burn Down the White House?' President Trump Fumbles in Phone Call With Justin Trudeau," *Time*, June 6, 2018, http://time.com/5304008/donald-trump-canada-tariffs-war-/.

14   For an amusing account of the United States' designs on Canada, see Kevin Lippert, *War Plan Red: The United States' Secret Plan to Invade Canada and Canada's Secret Plan to Invade the United States* (Princeton, NJ: Princeton Architectural Press, 2015), especially 58–68. Lippert also details a plan for Canada to invade the United States hatched after World War I by a disaffected military man that exists in only a few notes because it was ordered destroyed by military brass.

15   See my fictionalized biography of Nelson, *The Words on the Wall: Robert Nelson and the Rebellion of 1837* (Ottawa: Oberon Press, 1998).

16   John Boyko, "American Civil War and Canada," *Canadian Encyclopedia*, January 5, 2017, https://www.thecanadianencyclopedia.ca/en/article/american-civil-war.

17   David Stebenne, "Statehood for Puerto Rico? Lessons from the Last Time the U.S. Added a Star to Its Flag," *Huffington Post*, June 13, 2017, https://www.huffingtonpost.com/entry/statehood-for-puerto-rico-lessons-from-the-last-time_us_593bf3f6e4b014ae8c69e0c1. Stebenne also notes that in 1959 controversy swirled around the last two states admitted, Alaska and Hawaii, in part because Southern Democrats feared admitting multiracial Hawaii. In the end, largely white Alaska was admitted first as a compromise.

18   Vann R. Newkirk II, "The Situation in Puerto Rico Is Untenable," *Atlantic*, September 20, 2018, https://www.theatlantic.com/politics/archive/2018/09/hurricane-maria-anniversary-puerto-rico-trump/570928/.

19   Lippert, *War Plan Red*, 71–6 and 80–131. War Plan Red was only declassified in the 1970s.

20   Beth LaDow, *The Medicine Line: Life and Death on a North American Borderland* (New York: Routledge, 2002).

21   Marlene Leung, "A tale of two cities: Shared library, opera house lie on Canada-U.S. border," *CTV News*, May 11, 2015, http://www.ctvnews.ca/lifestyle/a-tale-of-two-cities-shared-library-opera-house-lie-on-canada-u-s-border-1.2365494.

22   John Boileau, "The story of the Maine-N.B. border," *Chronicle-Herald*, December 4, 2011.

23   Figures from https://www.worldatlas.com/articles/the-most-spoken-languages-in-america.html, and Statistics Canada, "Fast figures on Canada's official languages (2016)," available at https://www.clo-ocol.gc.ca/en/statistics/canada.

24   See Hyon B. Shin and Rosalind Bruno, "Language Use and English-Speaking Ability: 2000," US Census Bureau, October 2003, https://www.census.gov/prod/2003pubs/c2kbr-29.pdf, 4.

25   Calculated from US census data, available at: https://www2.census.gov/prod2/decennial/documents/33405927v1ch12.pdf, and https://www2.census.gov/prod2/decennial/documents/33405927v1ch02.pdf.

26   Camille Ryan, "Language Use in the United States: 2011," US Census Bureau, August 2013, https://www2.census.gov/library/publications/2013/acs/acs-22/acs-22.pdf.

See also Julia Preston, "The Truth About Mexican-Americans," *New York Review of Books*, December 3, 2015, http://www.nybooks.com/articles/2015/12/03/truth-about-mexican-americans/.

27 Jason Kaufman, *The Origins of Canadian and American Political Differences* (Cambridge, MA: Harvard University Press, 2009).

28 Ibid., 33–4.

29 See "Jacques Cartier," *Historical Narratives of Early Canada*, 2013, http://www.uppercanadahistory.ca/finna/finna1.html.

30 Bazilchuk and Strimbeck, *The Vermont Mountains*. See also, Lee Sultzman, "Abenaki History," July 21, 1997, http://www.tolatsga.org/aben.html.

31 See "The Great Peace of Montréal, 1701," Pointe-à-Callière, Musée d'archéologie et d'histoire de Montréal, 2012, http://www.virtualmuseum.ca/edu/ViewLoitCollection.do;jsessionid=036DF6215BA736792A99A30E4EB5E568?method=preview&lang=EN&id=24707.

32 Mark Cronlund Anderson, *Holy War: Cowboys, Indians, and 9/11s* (Regina: University of Regina Press, 2016). See especially 26–7.

33 Herman, *How the Scots Invented the Modern World*, 232.

34 Ibid, 233.

35 Thomas B. Edsall, "The Not-So-Silent White Majority," *New York Times*, November 17, 2016, http://www.nytimes.com/2016/11/17/opinion/the-not-so-silent-white-majority.html.

36 This difference has been constant for as long as the censuses of the two countries have been asking questions about Aboriginal/Indigenous ethnicity. See Statistics Canada, "Aboriginal Peoples in Canada: First Nations People, Métis and Inuit," 2011, https://www12.statcan.gc.ca/nhs-enm/2011/as-sa/99-011-x/99-011-x2011001-eng.cfm, and Stella U. Ogunwole, "The American Indian and Alaska Native Population: 2000," US Census Bureau, February 2002, https://www.census.gov/prod/2002pubs/c2kbr01-15.pdf.

37 "Aboriginal Racism in Canada," National Collaborating Centre for Aboriginal Health, 2019, http://www.nccah-ccnsa.ca/419/Aboriginal_Racism_in_Canada.nccah.

38 For an example, see Julia Page, "Quebec religious order apologizes, opens hotline for Innu victims of sexual abuse," *CBC News*, March 23, 2018, https://www.cbc.ca/news/canada/montreal/quebec-religious-order-apologizes-opens-hotline-for-innu-victims-of-sexual-abuse-1.4590085.

39 Mary Annette Pember, "When Will U.S. Apologize for Genocide of Indian Boarding Schools?" *Huffington Post*, December 6, 2017, https://www.huffingtonpost.com/mary-annette-pember/when-will-us-apologize-fo_b_7641656.html.

40 See NAACP, "History of Lynchings," 2019, http://www.naacp.org/history-of-lynchings/. Tellingly, the victim, an Indigenous man named Louie Sam, was killed by a mob of Americans who had crossed the border into Canada in 1884.

41 Natasha L. Henry, "Underground Railroad," *Canadian Encyclopedia*, March 4, 2015, https://www.thecanadianencyclopedia.ca/en/article/underground-railroad.

42 Anne Milan and Kelly Tran, "Blacks in Canada: A long history," *Canadian Social Trends* (2004), http://www.statcan.gc.ca/pub/11-008-x/2003004/article/6802-eng.pdf.

43 For the United States, see http://www.census.gov/quickfacts/table/PST045215/00; for Canada, see http://www12.statcan.gc.ca/nhs-enm/2011/dp-pd/prof/details/page.cfm?Lang=E&Geo1=PR&Code1=01&Data=Count&SearchText=Canada&SearchType=Begins&SearchPR=01&A1=All&B1=All&Custom=&TABID=1.

44 A.J. Somerset, "How Canadians Helped Create the NRA," *Toronto Star*, December 20, 2015, excerpt from *Arms: The Culture and Credo of the Gun* by A.J. Somerset (Windsor, ON: Biblioasis, 2015), https://www.thestar.com/news/insight/2015/12/20/how-canadians-helped-create-the-nra.html.

45 Max Roser, "Homicides," Our World in Data, 2019, https://ourworldindata.org/homicides.

46 Royal Canadian Mounted Police, "History of Firearms Control in Canada: Up to and Including the Firearms Act," September 2016, http://www.rcmp-grc.gc.ca/cfp-pcaf/pol-leg/hist/con-eng.htm.

47 It's worth quoting the poem in its entirety:
> In Flanders fields the poppies blow
> Between the crosses, row on row,
> That mark our place; and in the sky
> The larks, still bravely singing, fly
> Scarce heard amid the guns below.
>
> We are the Dead. Short days ago
> We lived, felt dawn, saw sunset glow,
> Loved, and were loved, and now we lie
> In Flanders fields.
>
> Take up our quarrel with the foe:
> To you from failing hands we throw
> The torch; be yours to hold it high.
> If ye break faith with us who die
> We shall not sleep, though poppies grow
> In Flanders fields.

Some helpful background is available on the Canadian War Museum's website, at http://www.warmuseum.ca/cwm/exhibitions/remember/flandersfields_e.shtml.

48 An excerpt from King's speech to the House of Commons is available at https://www.mta.ca/library/courage/w.l.mackenziekingdeclaresw.html. See also, Tim Cook, "How a wily Canadian prime minister entered WWII," *Toronto Star*, August 14, 2014, https://www.thestar.com/news/insight/2014/08/14/how_a_wily_canadian_prime_minister_entered_wwii.html.

49 See "Monroe Doctrine (1823)," *Ourdocuments.gov*, https://www.ourdocuments.gov/doc.php?flash=true&doc=23.

50 "The Nuclear Question in Canada," Diefenbaker Canada Centre, 2019, https://diefenbaker.usask.ca/virtual-exhibits/nuclear-question.php.

51 Lee Berthiaume, "Peacekeeping numbers reach 35-year low under Trudeau Liberals," *Global News*, October 23, 2017, https://globalnews.ca/news/3820090/peacekeeping-numbers-low-trudeau/.

52 Jeffrey Simpson, "Socialist policies will be history, Crowley predicts," reproduced at *Brian Lee Crowley* (blog), January 27, 2010, http://www.brianleecrowley.com/2010/01/jeffrey-simpson-reviews-fearful-symmetry/.

53 Political commentators wrote about this throughout Harper's reign. An example is Les Whittington, "2014 Canadian budget expected to keep 'starving the beast'," *Toronto Star*, January 17, 2014, https://www.thestar.com/news/canada/2014/01/17/2014_canadian_budget_expected_to_keep_starving_the_beast.html.

54 Government of Canada, "Canada's Health Care System," February 26, 2018, https://www.canada.ca/en/health-canada/services/health-care-system/reports-publications/health-care-system/canada.html#a13.

55 Krugman, *Conscience*, especially 3–14.

56 See "Elections in Canada," *Canada Guide*, 2019, http://www.thecanadaguide.com/government/elections/, and Federal Election Commission, "Contribution limits," https://www.fec.gov/help-candidates-and-committees/candidate-taking-receipts/contribution-limits/.

57 Richard V. Reeves and Pete Rodrigue, "Has the American Dream Moved to Canada?" *Brookings*, July 1, 2014, https://www.brookings.edu/blog/social-mobility-memos/2014/07/01/has-the-american-dream-moved-to-canada/.

58 Marie Allard, "Sophie Grégoire-Trudeau et le nom des femmes," *La Presse*, January 23, 2016, http://www.lapresse.ca/vivre/societe/201601/22/01-4942634-sophie-gregoire-trudeau-et-le-nom-des-femmes.php.

59 Fifty-eight percent of Quebec parents reported using day care in the preceding year, according to a 2013 survey; the rate was the highest in Canada. See Maire Sinha, "Child care in Canada," available at https://www150.statcan.gc.ca/n1/pub/89-652-x/89-652-x2014005-eng.htm#n7-refa.

60 Rachelle Younglai, "Canadian women more active in labour force than American peers," *Globe and Mail*, August 17, 2016, http://www.theglobeandmail.com/report-on-business/economy/canadian-womens-work-force-participation-surges/article31450579/. See also Government of Canada, "EI Maternity and Parental Benefits—Overview," available at https://www.canada.ca/en/services/benefits/ei/ei-maternity-parental.html.

61 Alexia Fernández Campbell, "Americans want paid parental leave. The Rubio-Ivanka plan is a sorry attempt," *Vox*, August 6, 2018, https://www.vox.com/2018/8/6/17648462/rubio-ivanka-republican-paid-leave.

62 Lisa Tucker, "California Maternity Leave: Everything Expectant Parents Need to Know," *Working Mother*, January 10, 2018, https://www.workingmother.com/california-maternity-leave-everything-expectant-parents-need-to-know.

63 Statistics Canada, "Table W21-29 Summary of total full-time teachers, by level of instruction, Canada, 1960 to 1975," available at https://www150.statcan.gc.ca/n1/pub/11-516-x/sectionw/4147445-eng.htm.

64 Brooke Anderson, "The elephant in the (class)room: The debate over Americanization of Canadian universities and the question of national identity," *Studies by Undergraduate Researchers at Guelph* 4., no. 2 (2011), https://journal.lib.uoguelph.ca/index.php/surg/article/view/1315.

65 Ibbitson, *Stephen Harper*, 28.

66 Fred Langan, "Prolific Musician Fell in Love with Canada," *Globe and Mail*, April 15, 2014, https://www.theglobeandmail.com/arts/music/prolific-musician-fell-in-love-with-canada/article17999477/.

67 Development of Quebec's hydroelectric potential was particularly important. Among the spinoffs were the establishment of Francophone engineering firms, the promise of low electricity rates for the foreseeable future in order to stimulate industry, and the requirement that firms deal with the government in French. For more on the Quiet Revolution and the beginning of Quebec, Inc., see "1930–1944—Toward Nationalization," available at http://www.hydroquebec.com/history-electricity-in-quebec/timeline/toward-nationalization.html.

68 See Brent Patterson, "The false claim of the USMCA's cultural protections," *Rabble*, October 15, 2018, http://rabble.ca/blogs/bloggers/brent-patterson/2018/10/false-claim-usmcas-cultural-protections, and Michael Geist, "The Full 'Culture Exception' That Isn't: Why Canada Caved on Independent Cultural Policy in the USMCA," *Michael Geist* (blog), October 11, 2018, http://www.michaelgeist.ca/2018/10/usmcaculture/.

69 The only radio voices in the States that are heard all over the country are those of conservative commentators like Russ Limbaugh. National Public Broadcasting stations have far less coverage and smaller audiences. And this is not to mention the reach of Fox News on television, which many blame for being far too soft on the Republicans during the 2016 election, and even more friendly to Donald Trump after he became president. See Bill Mann, "Here's Why the Right Wing Dominates Talk Radio Today," *HuffPost*, May 25, 2011, http://www.huffingtonpost.com/bill-mann/heres-why-the-right-wing_b_206444.html.

70 Figures from Friends of Canadian Broadcasting, "Change in Parliamentary Appropriation to the CBC (in 2014 $)," April 10, 2014, https://friends.ca/explore/article/change-in-parliamentary-appropriation-to-cbc-in-2014.

71 Andy Horwitz, "Who Should Pay for the Arts in America?" *Atlantic*, January 31, 2016, https://www.theatlantic.com/entertainment/archive/2016/01/the-state-of-public-funding-for-the-arts-in-america/424056/.

72 She was eliminated from the race on the ninth of thirteen ballots; see "Scheer wins Conservative leadership race: results by ballot," *Hill Times*, May 27, 2017, https://www.hilltimes.com/2017/05/27/conservative-leadership-race-results-ballot/108523. See also Anne Kingston, "How Kellie Leitch accidentally revealed

Canadian values," *Maclean's*, May 26, 2017, http://www.macleans.ca/politics/how-kellie-leitch-accidentally-revealed-canadian-values/.

73 Cecilia Keating, " 'I'm not perfect': CAQ leader softens tone on immigration in final Quebec election debate," *National Observer*, September 20, 2018, https://www.nationalobserver.com/2018/09/20/news/im-not-perfect-caq-leader-softens-tone-immigration-final-quebec-election-debate.

74 See "2018 Midterm Election Results: Live," *New York Times*, January 28, 2019, https://www.nytimes.com/interactive/2018/11/06/us/elections/results-dashboard-live.html.

75 Calculated from Simon Fraser University's data sets, available at https://www.sfu.ca/~aheard/elections/results.html.

76 Canadian Press, "NDP struggles to raise money as rival parties boast of record fundraising hauls," *Toronto Star*, January 30, 2019, https://www.thestar.com/news/canada/2019/01/30/ndp-struggles-to-raise-money-as-rival-parties-boast-of-record-fundraising-hauls.html.

77 "Ontario election results 2018: A map of the results," *Globe and Mail*, June 8, 2018, https://www.theglobeandmail.com/canada/article-ontario-election-results-2018-a-map-of-the-live-results/.

78 Elections Quebec figures available at https://www.electionsquebec.qc.ca/provinciales/en/results.php.

79 Figures available at http://www.cnn.com/election/results/president.

80 Final popular congressional vote: Democrats, 60,727,598 votes; Republicans, 50,983,895 votes. For Senate seats: Democrats, 53,078,387 votes; Republicans, 34,981,675 votes. Figures from "2018 House Popular Vote Tracker," available at https://docs.google.com/spreadsheets/d/1WxDaxD5az6kdOjJncmGph37z0BPNhV1fNAH_g7IkpC0/edit#gid=0, and "U.S. Senate Election Results 2018," *New York Times*, January 28, 2019, https://www.nytimes.com/interactive/2018/11/06/us/elections/results-senate-elections.html.

81 "Canada's immigration website crashes on election night," CTV *News*, November 9, 2016, http://www.ctvnews.ca/canada/canada-s-immigration-website-crashes-on-election-night-1.3152231.

82 Patrick Cain, "About 2,000 more Americans than normal have moved to Canada since Trump's election," *Global News*, August 21, 2018, https://globalnews.ca/news/4396938/move-to-canada-donald-trump/.

83 For a brief history of the highs and the lows of Canada's behaviour toward refugees, see the Canadian Council for Refugees website, at http://ccrweb.ca/en/brief-history-canadas-responses-refugees.

84 See Monica Boyd and Michael Vickers, "100 years of immigration in Canada," *Canadian Social Trends* (2000), https://pdfs.semanticscholar.org/2639/99b35215839a9445829905f0322657a84ed3.pdf, and Gustavo López, Kristen Bialik, and Jynnah Radford, "Key findings about U.S. immigrants," *Pew Research Center*, September 14, 2018, http://www.pewresearch.org/fact-tank/2018/09/14/key-findings-about-u-s-immigrants/.

85 Government of Canada, "Notice—Supplementary Information 2018–2020 Immigration Levels Plan," November 1, 2017, https://www.canada.ca/en/immigration-refugees-citizenship/news/notices/supplementary-immigration-levels-2018.html.

86 See Statistics Canada, "150 years of immigration in Canada," June 29, 2016, https://www150.statcan.gc.ca/n1/pub/11-630-x/11-630-x2016006-eng.htm.

87 National Conference of State Legislatures, "Snapshot of U.S. Immigration 2017," August 3, 2017, http://www.ncsl.org/research/immigration/snapshot-of-u-s-immigration-2017.aspx.

88 So porous was the Canada-US border that an aunt, born in Saskatchewan in 1913 and who entered the United States when her parents returned about 1916, did not realize that she wasn't an American citizen until she applied for a US passport in her fifties. She ended up having to be naturalized.

89 Xiaojian Zhao, "Immigration to the United States after 1945," *Oxford Research Encyclopedias*, July 2016, http://americanhistory.oxfordre.com/view/10.1093/acrefore/9780199329175.001.0001/acrefore-9780199329175-e-72.

90 Editorial Board, "U.S. Farms Can't Compete Without Foreign Workers," *Bloomberg*, June 1, 2018, https://www.bloomberg.com/view/articles/2018-06-01/u-s-farms-need-more-immigrant-workers.

91 Adrian Morrow, "Trump points to Canada as a model for U.S. immigration reform in Congress speech," *Globe and Mail*, February 28, 2017, https://www.theglobeandmail.com/news/world/us-politics/donald-trump-congress-speech-immigration/article34158135/.

92 Watch Sarah Huckabee Sanders's comments at https://www.youtube.com/watch?v=-V8QM7UZQTQ.

93 Z. Byron Wolf, "Trump blasts 'breeding' in sanctuary cities. That's a racist term," CNN, April 24, 2018, https://www.cnn.com/2018/04/18/politics/donald-trump-immigrants-california/index.html.

94 Derek Thompson, "Canada's Secret to Escaping the 'Liberal Doom Loop,'" *Atlantic*, July 9, 2018, https://www.theatlantic.com/ideas/archive/2018/07/canadas-secret-to-escaping-the-liberal-doom-loop/564551/.

95 Adam Nagourney, "A Democratic Rout in Orange County: Cisneros's Win Makes It Four" *New York Times*, November 17, 2018, https://www.nytimes.com/2018/11/17/us/politics/cisneros-orange-county-democrats.html.

96 Viet Thanh Nguyen, "Could Asian-Americans Turn Orange County Blue?" *New York Times*, November 5, 2018, https://www.nytimes.com/2018/11/05/opinion/asian-americans-orange-county-house-races.html.

97 Giuseppe Valiante, "U.S. Vietnam war draft dodgers left their mark on Canada," *Maclean's*, April 16, 2015, https://www.macleans.ca/news/canada/u-s-vietnam-war-draft-dodgers-left-their-mark-on-canada/.

98 Australia, in much closer proximity to the post–Vietnam War humanitarian crisis, admitted the most proportionately. In 2017, the US Census Bureau

estimated that 2.1 million people of Vietnamese descent (about 0.7 percent of the US population) lived in the United States, while in Canada the figure in 2016 was 240,615 (also about 0.7 percent). In Australia in 2016 the figures were 219,357, or about 0.9 percent. These figures are available at https://factfinder.census.gov/faces/tableservices/jsf/pages/productview.xhtml?pid=ACS_17_1YR_B02018&prodType=table; https://www12.statcan.gc.ca/census-recensement/2016/dp-pd/prof/details/page.cfm?Lang=E&Geo1=PR&Code1=01&Geo2=PR&Code2=01&Data=Count&SearchText=canada&SearchType=Begins&SearchPR=01&B1=Ethnic%20origin&TABID=1; http://www.abs.gov.au/ausstats/abs@.nsf/2f762f95845417aeca25706c00834efa/e0a8b4f57a46da56ca2570ec007853c9!OpenDocument.

99 For those who would like to see how writers of the Vietnamese diaspora deal with their heritage, three prize-winning novels are instructive. Two are by Canadian-Vietnamese authors: *Ru* by Kim Thúy (trans. Sheila Fischman; Toronto: Random House, 2012), which in poetic French tells of a woman born during the Tết Offensive in 1968 who comes to Canada as a boat person. The second is Vincent Lam's *The Headmaster's Wager* (Toronto: Doubleday Canada, 2012). It takes place during the Vietnam War itself, and ends with the protagonist's young grandson leaving at the fall of Saigon. *The Sympathizer* (New York: Grove Press, 2015) by Viet Thanh Nguyen is the third. It centres on what happens after the war to a young Vietnamese man who finds himself in the United States among people for whom the war is not over, even twenty years later, and who would like to go back and replay history so the forces of the South win.

100 Brett J. Skinner, "Medicare, the Medical Brain Drain and Human Resource Shortages in Health Care," *AIMS Health Care Reform Background Paper #7* (December 2002), http://www.aims.ca/site/media/aims/BrainDrain.pdf.

101 Shefali Luthra, "Heading North: American Doctors Report Back From Canada," *NPR*, December 14, 2017, https://www.npr.org/sections/health-shots/2017/12/14/570216320/heading-north-american-doctors-report-back-from-canada.

102 While 83 percent of Quebecers told Statistics Canada they were Roman Catholic in 2001, attendance at Roman Catholic churches has been in freefall for the last forty years.

103 "La communauté vietnamienne fait revivre une église de Montréal," *Radio-Canada*, September 7, 2015, http://ici.radio-canada.ca/emissions/le_15_18/2015-2016/chronique.asp?idChronique=382451.

TWINS: THE LAST WORDS

1   The incidence of fraternal twins varies greatly: they account for about 6 in 1,000
     births in Japan and as many as 20 in 1,000 in parts of Africa. They are no more
     alike than ordinary brothers and sisters would be. See "Fraternal Twins vs.
     Identical Twins," available at http://www.diffen.com/difference/Fraternal_Twins_
     vs_Identical_Twins.

SELECTED **BIBLIOGRAPHY**

Abbas, Samar. "India's Parthian Colony." Circle of Ancient Iranian Studies. http://www.cais-soas.com/CAIS/History/ashkanian/parthian_colony.htm.

Agence France-Presse. "Rwanda: non-lieu requis en France dans l'enquête sur l'attentat déclencheur du génocide." *L'Express*, October 13, 2018. https://www.lexpress.fr/actualites/1/societe/rwanda-non-lieu-requis-en-france-dans-l-enquete-sur-l-attentat-declencheur-du-genocide_2039706.html.

Alchon, Suzanne Austin. *A Pest in the Land: New World Epidemics in a Global Perspective*. Albuquerque: University of New Mexico Press, 2003.

Allard, Marie. "Sophie Grégoire-Trudeau et le nom des femmes." *La Presse*, January 23, 2016. http://www.lapresse.ca/vivre/societe/201601/22/01-4942634-sophie-gregoire-trudeau-et-le-nom-des-femmes.php.

Amnesty International. *"Where Are We Going to Live?" Migration and Statelessness in Haiti and the Dominican Republic*. London: Amnesty International, 2016. https://www.amnesty.org/en/documents/amr36/4105/2016/en/.

———. "Rwanda: Decades of attacks repression and killings set the scene for next month's election." July 7, 2017. https://www.amnesty.org/en/latest/news/2017/07/rwanda-decades-of-attacks-repression-and-killings-set-the-scene-for-next-months-election/.

Anderson, Brooke. "The elephant in the (class)room: The debate over Americanization of Canadian universities and the question of national identity." *Studies by Undergraduate Researchers at Guelph* 4, no. 2 (2011). https://journal.lib.uoguelph.ca/index.php/surg/article/view/1315.

Anderson, Mark Cronlund. *Holy War: Cowboys, Indians, and 9/11s*. Regina: University of Regina Press, 2016.

Anderson, Stephen R. "How Many Languages Are There in the World?" Linguistic Society of America. http://www.linguisticsociety.org/content/how-many-languages-are-there-world.

André, Richard. "The Dominican Republic and Haiti: A Shared View from the Diaspora. A conversation with Edwidge Danticat and Junot Díaz." *Americas Quarterly* (Summer 2014). http://americasquarterly.org/content/dominican-republic-and-haiti-shared-view-diaspora.

Andrei, Mihai. "Doggerland—the land that connected Europe and the UK 8,000 years ago." zme *Science*, February 16, 2017. http://www.zmescience.com/science/geology/doggerland-europe-land/.

Anker, Lane. "Peacekeeping and Public Opinion." *Canadian Military Journal*, June 20, 2005. http://www.journal.forces.gc.ca/v06/n02/public-eng.asp.

Aristotle. *Politics*. Translated by J.E.C. Welldon. New York: Macmillan, 1893. https://archive.org/stream/in.ernet.dli.2015.216306/2015.216306.The-Political_djvu.txt.

Barber, Elizabeth Wayland. *The Mummies of Ürümchi*. New York: W.W. Norton & Company, 2004.

Barsh, Gregory S. "What Controls Variation in Human Skin Color?" PLOS *Biology* 1, no. 1 (2003). https://doi.org/10.1371/journal.pbio.0000027.

Bayon, Germain, Bernard Dennielou, Joël Etoubleau, Emmanuel Ponzevera, Samuel Toucanne, and Sylvain Bermell. "Intensifying Weathering and Land Use in Iron Age Central Africa." *Science* 335, no. 6073 (2012). http://dx.doi.org/10.1126/science.1215400.

Bazilchuk, Nancy, and Rick Strimbeck. *Longstreet Highroad Guide to the Vermont Mountains*. Atlanta: Longstreet Press, 1997.

Bellesiles, Michael A. *Revolutionary Outlaws: Ethan Allen and the Struggle for Independence on the Early American Frontier*. Charlottesville: University Press of Virginia, 1993.

Bello, Silvia M., Rosalind Wallduck, Simon A. Parfitt, and Chris B. Stringer. "An Upper Palaeolithic engraved human bone associated with ritualistic cannibalism." PLOS ONE 12, no. 8 (2017). https://doi.org/10.1371/journal.pone.0182127.

Blake, Jonathan S. "What a Protestant Parade Reveals About Theresa May's New Partners." *Atlantic*, July 11, 2017. https://www.theatlantic.com/international/archive/2017/07/protestant-parade-northern-ireland/533151/.

Boileau, John. "The story of the Maine-N.B. border." *Chronicle-Herald*, December 4, 2011.

Boyd, Monica, and Michael Vickers. "100 years of immigration in Canada." *Canadian Social Trends* (2000). https://pdfs.semanticscholar.org/2639/99b35215839a9445829905f0322657a84ed3.pdf.

Bozonnet, Charlotte. "En Algérie, 'il reste beaucoup à faire' pour l'égalité des femmes." *Le Monde*, March 13, 2015. http://www.lemonde.fr/afrique/article/2015/03/13/en-algerie-il-reste-beaucoup-a-faire-pour-l-egalite-des-femmes_4593413_3212.html#tmktHUhhjRJHxD2Y.99.

Bown, Stephen R. *1494: How a Family Feud in Medieval Spain Divided the World in Half*. Vancouver: Douglas & McIntyre, 2011.

Caesar, Julius. *The Gallic Wars.* Translated by W.A. McDevitte and W.S. Bohn. Internet Classics Archive, 1994–2000. http://classics.mit.edu/Caesar/gallic.html.

Cahill, Thomas. *How the Irish Saved Civilization: The Untold Story of Ireland's Heroic Role from the Fall of Rome to the Rise of Medieval Europe.* New York: Doubleday, 1995.

Caldwell, Robert. *A comparative grammar of the Dravidian, or, South-Indian family of languages.* London: Trubner & Co., 1875.

Camps, Gabriel. "Tableau chronologique de la Préhistoire récente du Nord de l'Afrique." *Bulletin de la Société préhistorique française.* Études et travaux 65, no. 2 (1968): 609–22.

Cassidy, Lara M., Rui Martiniano, Eileen M. Murphy, Matthew D. Teasdale, James Mallory, Barrie Hartwell, and Daniel G. Bradley. "Neolithic and Bronze Age migration to Ireland and establishment of the insular Atlantic genome." *PNAS* 113, no. 2 (2016). https://doi.org/10.1073/pnas.1518445113.

Chardy, Alfonso. "How Fidel Castro and the Mariel boatlift changed lives and changed Miami." *Miami Herald,* November 26, 2016. http://www.miamiherald.com/news/nation-world/world/americas/fidel-castro-en/article117206643.html.

Charrad, Mounira M. *States and Women's Rights: The Making of Postcolonial Tunisia, Algeria, and Morocco.* Berkeley and Los Angeles: University of California Press, 2001.

Chrétien, Jean-Pierre. *L'Afrique des Grands Lacs: Deux mille ans d'histoire.* Paris: Aubier, 2000.

Coedès, Georges. *Les Peuples de la Péninsule Indochinoise: Histoire—Civilisations.* Paris: Dunod, 1962.

Conway, Aidan, and J.F. Conway. "Saskatchewan: From Cradle of Social Democracy to Neoliberalism's Sandbox." In *Transforming Provincial Politics: The Political Economy of Canada's Provinces and Territories in the Neoliberal Era,* edited by Bryan M. Evans and Charles W. Smith, 226–54. Toronto: University of Toronto Press, 2015.

Danino, Michel. "Genetics and the Aryan Debate." *Puratattva: Bulletin of the Indian Archaeological Society* 36 (2005–6): 146–54.

Danticat, Edwidge. *The Farming of Bones.* New York: Soho Press, 1998.

Daoud, Kamel. "The Algerian Exception." *New York Times,* May 29, 2015. http://www.nytimes.com/2015/05/30/opinion/the-algerian-exception.html?_r=0.

Diamond, Jared. *Collapse: How Societies Choose to Fail or Succeed.* New York: Viking Press, 2005.

———. *Guns, Germs, and Steel: The Fates of Human Societies.* New York and London: W.W. Norton and Company, 1999.

Dio, Cassius. *Roman History.* Translation by Earnest Cary. Cambridge, MA: Harvard University Press, 1914. http://penelope.uchicago.edu/Thayer/e/roman/texts/cassius_dio/home.html

Doty, Robert C. "Riots Dim Algeria Peace Hopes." *New York Times,* October 22, 1961.

Dubois, Laurent. *Haiti: The Aftershocks of History.* New York: Picador, 2003.

Duffy, Judith. "Welcome to Secular Scotland…a nation where religion is in retreat." *Herald*, May 29, 2016. http://www.heraldscotland.com/news/14523231.Welcome_to_Secular_Scotland_____a_nation_where_religion_is_in_retreat.

Durland, William Davies. "The Forests of the Dominican Republic." *Geographical Review* 12, no. 2 (1922): 206–22.

Eiberg, Hans, Jesper Troelsen, Mette Nielsen, Annemette Mikkelsen, Jonas Mengel-From, Klaus W. Kjaer, and Lars Hansen. "Blue eye color in humans may be caused by a perfectly associated founder mutation in a regulatory element located within the HERC2 gene inhibiting OCA2 expression." *Human Genetics* 123, no. 2 (2008). https://doi.org/10.1007/s00439-007-0460-x.

The Economist. "Southern comfort: Tamil Nadu and Kerala dance to a different tune from the rest of India." *Economist*, May 28, 2016. https://www.economist.com/asia/2016/05/28/southern-comfort.

Fall, Bernard B. *Street Without Joy*. Harrisburg, PA: The Stackpole Company, 1967.

Faye, Gaël. *Petit Pays*. Paris: French and European Publications, 2017.

Ferguson, Will. *Road Trip Rwanda: A Journey into the New Heart of Africa*. Toronto: Viking, 2005.

Fernández Campbell, Alexia. "Americans want paid parental leave. The Rubio-Ivanka plan is a sorry attempt." *Vox*, August 6, 2018. https://www.vox.com/2018/8/6/17648462/rubio-ivanka-republican-paid-leave.

Gonin, Jean-Marc. "Les pieds-noirs, 50 ans après." *Le Figaro*, October 8, 2012. http://www.lefigaro.fr/actualite-france/2012/01/27/01016-20120127ARTFIG00422-les-pieds-noirs-50-ans-apres.php.

Gozlan, Martine. *Tunisie-Algérie-Maroc: la colère des peoples*. Paris: L'Archipel, 2011.

Graham, John W. *Whose Man in Havana? Adventures from the Far Side of Diplomacy*. Calgary: University of Calgary Press, 2015.

Grant, R.G., Ann Kay, Michael Kerrigan, and Philip Parker. *History of Britain and Ireland: The Definitive Visual Guide*. New York: DK Publishers, 2011.

Grant, Richard. "Paul Kagame: Rwanda's redeemer or ruthless dictator?" *Telegraph*, July 22, 2010. http://www.telegraph.co.uk/news/worldnews/africaandindianocean/rwanda/7900680/Paul-Kagame-Rwandas-redeemer-or-ruthless-dictator.html.

Gunter, Joel. "Abortion in Ireland: The fight for choice." *BBC News*, March 8, 2017. http://www.bbc.com/news/world-europe-39183423.

Haggerty, Richard A., ed. *Dominican Republic: A Country Study*. Washington, DC: GPO for the Library of Congress, 1989. http://countrystudies.us/dominican-republic/4.htm.

——. *Haiti: A Country Study*. Washington, DC: GPO for the Library of Congress, 1989. http://countrystudies.us/haiti/8.htm.

Harrison, Blake. *The View from Vermont: Tourism and the Making of an American Rural Landscape*. Burlington, VT: University of Vermont Press, 2006.

Hayes, Derek. "Drawing the lines." *Canadian Geographic* 125, no. 1 (2005): 48–9.

Hayes, Jeffrey. "Women in Vietnam: Traditional Views, Advances and Abuse." *Facts and Details* (blog), May 2014. http://factsanddetails.com/southeast-asia/Vietnam/sub5_9c/entry-3390.html#chapter-2Vietnamese.

Herodotus. *The Histories*. Translated by Aubrey de Sélincourt. London: Penguin, 1972.

Horwitz, Andy. "Who Should Pay for the Arts in America?" *Atlantic*, January 31, 2016. https://www.theatlantic.com/entertainment/archive/2016/01/the-state-of-public-funding-for-the-arts-in-america/424056.

Hublin, Jean-Jacques, Abdelouahed Ben-Ncer, Shara E. Bailey, Sarah E. Freidline, Simon Neubauer, Matthew M. Skinner, Inga Bergmann, Adeline Le Cabec, Stefano Benazzi, Katerina Harvati, and Philipp Gunz. "New fossils from Jebel Irhoud, Morocco and the pan-African origin of *Homo sapiens*." *Nature* 546, no. 7657 (2017). https://www.nature.com/articles/nature22336.

Ibbitson, John. *Stephen Harper*. Toronto: Signal, 2015.

International Commission of Inquiry for Burundi. *International Commission of Inquiry for Burundi: Final Report*. 2004. https://www.usip.org/sites/default/files/file/resources/collections/commissions/Burundi-Report.pdf.

International Federation for Human Rights. *Burundi on the brink, looking back on two years of terror*. Report No. 693a, June 2017. https://www.fidh.org/IMG/pdf/burundi_jointreport_june2017_eng_final.pdf.

Ireland, Corydon. "Vermont and New Hampshire, geographic twins, cultural aliens." *Harvard Gazette*, November 1, 2007. http://news.harvard.edu/gazette/story/2007/11/vermont-and-new-hampshire-geographic-twins-cultural-aliens/.

Jackes, Mary, and David Lubell. "Early and Middle Holocene Environments and Capsian Cultural Change: Evidence from the Télidjène Basin, Eastern Algeria." *African Archaeological Review* 25, no. 1–2 (2008): 41–55. https://doi.org/10.1007/s10437-008-9024-2.

Joseph, Manu. "Setting a High Bar for Poverty in India." *New York Times*, July 9, 2014. http://www.nytimes.com/2014/07/10/world/asia/setting-a-high-bar-for-poverty-in-india.html.

Kagire, Edmund. "Rwanda's Mushikiwabo takes the reins at Francophone club of 58 countries." *East African*, October 13, 2018. http://www.theeastafrican.co.ke/news/ea/Rwanda-Mushikiwabo-takes-the-reins-at-Francophonie/4552908-4803792-12l17j6z/index.html.

Kamen, Henry. *Spain's Road to Empire: The Making of a World Power, 1492–1763*. London: Penguin, 2009.

Kaufman, Jason. *The Origins of Canadian and American Political Differences*. Cambridge, MA: Harvard University Press, 2009.

Keay, John. *India: A History*. New York: Harper Collins, 2000.

Keys, David. "Kingdom of the Sands." *Archeology* 57, no. 2 (2004). http://archive.archaeology.org/0403/abstracts/sands.html.

———. "The lost empire explored: The Cholas once had great power, but the world has forgotten them." *Independent*, May 9, 1993. http://www.independent.co.uk/

arts-entertainment/travel-the-lost-empire-explored-the-cholas-once-had-great-power-but-the-world-has-forgotten-them-2321900.html.

Kiernan, Ben. *Blood and Soil: A World History of Genocide and Extermination from Sparta to Darfur.* New Haven, CT: Yale University Press, 2007.

———. "The First Genocide: Carthage, 146 BC." *Diogenes* 51, no. 3 (2004): 27–39. https://journals.sagepub.com/doi/10.1177/0392192104043648.

Kilbourne, Frederick W. *Chronicles of the White Mountains.* Boston and New York: Houghton Mifflin, 1916.

Kingston, Anne. "How Kellie Leitch accidentally revealed Canadian values." *Maclean's,* May 26, 2017. http://www.macleans.ca/politics/how-kellie-leitch-accidentally-revealed-canadian-values/.

Krugman, Paul. *The Conscience of a Liberal.* New York: W.W. Norton & Company, 2007.

LaDow, Beth. *The Medicine Line: Life and Death on a North American Borderland.* New York: Routledge, 2002.

Lageman, Thessa. "Mohamed Bouazizi: Was the Arab Spring worth dying for?" *Al Jazeera,* January 3, 2016. http://www.aljazeera.com/news/2015/12/mohamed-bouazizi-arab-spring-worth-dying-151228093743375.html.

Lemarchand, René. *Burundi: Ethnic Conflict and Genocide.* Cambridge and New York: Woodrow Wilson Center Press and Cambridge University Press, 1996.

Leonardi, Michela, Pascale Gerbault, Mark G. Thomas, and Joachim Burger. "The evolution of lactase persistence in Europe. A synthesis of archaeological and genetic evidence." *International Dairy Journal* 22, no. 2 (2012). https://dx.doi.org/10.1016%2Fj.idairyj.2011.10.010.

Leung, Marlene. "A tale of two cities: Shared library, opera house lie on Canada-U.S. border." *CTV News,* May 11, 2015. http://www.ctvnews.ca/lifestyle/a-tale-of-two-cities-shared-library-opera-house-lie-on-canada-u-s-border-1.2365494.

Li, Chunxiang, Chao Ning, Erika Hagelberg, Hongjie Li, Yongbin Zhao, Wenying Li, Idelisi Abuduresule, Hong Zhu, and Hui Zhou. "Analysis of ancient human mitochondrial DNA from the Xiaohe cemetery: Insights into prehistoric population movements in the Tarim Basin, China." *BMC Genetics* 16, no. 78 (2015). https://doi.org/10.1186/s12863-015-0237-5.

Lippert, Kevin. *War Plan Red: The United States' Secret Plan to Invade Canada and Canada's Secret Plan to Invade the United States.* Princeton, NJ: Princeton Architectural Press, 2015.

Lipset, Seymour Martin. *Agrarian Socialism: The Cooperative Commonwealth Federation in Saskatchewan.* Berkeley: University of California Press, 1971.

Long, Kat. "Stephen Harper's Franklin fever." *National Post,* May 22, 2014. http://nationalpost.com/opinion/kat-long-stephen-harpers-franklin-fever/wcm/f68dc0b2-e595-4745-a09c-dc7d5d235a04.

López, Gustavo, Kristen Bialik, and Jynnah Radford. "Key findings about U.S. immigrants." *Pew Research Center,* September 14, 2018. http://www.pewresearch.org/fact-tank/2018/09/14/key-findings-about-u-s-immigrants/.

Lothrop, Jonathan C., Paige E. Newby, Arthur E. Spiess, and James W. Bradley. "Paleoindians and the Younger Dryas in the New England-Maritimes Region." *Quaternary International* 242, no. 2 (2011). https://doi.org/10.1016/j.quaint.2011.04.015.

Lynch, John. *Simón Bolívar: A Life*. New Haven, CT: Yale University Press, 2006.

Macpherson, C.B. *Democracy in Alberta: Social Credit and the Party System*. Toronto: University of Toronto Press, 1953.

Maddison, Angus. *The World Economy: A Millennial Perspective*. OECD Publishing, 2001.

Mallick, Chandana Basu, Florin Mircea Iliescu, Märt Möls, Sarah Hill, Rakesh Tamang, Gyaneshwer Chaubey, Rie Goto, Simon Y.W. Ho, Irene Gallego Romero, Federica Crivellaro, Georgi Hudjashov, Niraj Rai, Mait Metspalu, C.G. Nicholas Mascie-Taylor, Ramasamy Pitchappan, Lalji Singh, Marta Mirazon-Lahr, Kumarasamy Thangaraj, Richard Villems, and Toomas Kivisild. "The Light Skin Allele of SLC24A5 in South Asians and Europeans Shares Identity by Descent." *PLOS Genet* 9, no. 11 (2013). https://doi.org/10.1371/journal.pgen.1003912.

Mattingly, D.J., and M. Sterry. "The first towns in the central Sahara." *Antiquity* 87, no. 336 (2013): 503–18. https://doi.org/10.1017/S0003598X00049097.

McDonald, Henry. "Ireland's first gay prime minister Leo Varadkar formally elected." *Guardian*, June 14, 2017. https://www.theguardian.com/world/2017/jun/14/leo-varadkar-formally-elected-as-prime-minister-of-ireland.

McGoogan, Ken. *Celtic Lightning: How the Scots and the Irish Created a Canadian Nation*. Toronto: Patrick Crean Editions, 2015.

McKirdy, Euan. "Scottish referendum in doubt after steep losses for SNP in UK vote." *CNN*, June 9, 2017. http://www.cnn.com/2017/06/08/europe/snp-uk-general-election/index.html.

Meredith, Robyn. "5 Days in 1967 Still Shake Detroit." *New York Times*, July 23, 1997. http://www.nytimes.com/1997/07/23/us/5-days-in-1967-still-shake-detroit.html?pagewanted=all.

Miles, Richard. *Carthage Must Be Destroyed: The Rise and Fall of an Ancient Civilization*. New York: Viking, 2010.

Mills, Sean. *A Place in the Sun: Haiti, Haitians, and the Remaking of Quebec*. Montreal: McGill-Queen's University Press, 2016.

Morrow, Adrian. "Trump points to Canada as a model for U.S. immigration reform in Congress speech." *Globe and Mail*, February 28, 2017. https://www.theglobeandmail.com/news/world/us-politics/donald-trump-congress-speech-immigration/article34158135/.

Murphy, Dervla. *On a Shoestring to Coorg: An Experience of Southern India*. London: Flamingo, 1995.

Newman, Stanley. "Morris Swadesh." *Language* 43, no. 4 (1967): 948–57.

Nguyen, Viet Thanh. "Could Asian-Americans Turn Orange County Blue?" *New York Times*, November 5, 2018. https://www.nytimes.com/2018/11/05/opinion/asian-americans-orange-county-house-races.html.

———. *The Sympathizer*. New York: Grove Press, 2015.

Norwich, John Julius. *The Middle Sea: A History of the Mediterranean*. London: Vintage Books, 2007.

Ostrer, Harry, and Karl Skorecki. "The population genetics of the Jewish people." *Human Genetics* 132, no. 2 (2013): 119–27. https://doi.org/10.1007/s00439-012-1235-6.

Parker, George, Jim Pickard, and Arthur Beesley. "Theresa May wins Queen's Speech vote with slender majority." *Financial Times*, June 29, 2017. https://www.ft.com/content/6ddfc7a0-5cd7-11e7-b553-e2df1b0c3220?mhq5j=e3.

Payne, Stanley G. *A History of Spain and Portugal*, vol. 1. Madison: University of Wisconsin Press, 1973.

Pember, Mary Annette. "When Will U.S. Apologize for Genocide of Indian Boarding Schools?" *Huffington Post*, December 6, 2017. https://www.huffingtonpost.com/mary-annette-pember/when-will-us-apologize-fo_b_7641656.html.

Pereira Martins, Antonio Carlos. "Ensino superior no Brasil: da descoberta aos dias atuais." *Acta Cirurgica Brasileira* 17, no. 3 (2002). http://dx.doi.org/10.1590/S0102-86502002000900001.

Peterson, Mark A. *The Price of Redemption: The Spiritual Economy of Puritan New England*. Palo Alto, CA: Stanford University Press, 1997.

Pfaff, Alexander. "From Deforestation to Reforestation in New England, United States." In *World Forests from Deforestation to Transition? World Forests*, vol. 2, edited by M. Palo and H. Vanhanen, 67–82. Dordrecht, NL: Springer, 2000.

Polo, Marco. *The Travels*. Translated and with an introduction by Ronald Latham. London: Penguin Classics, 1958.

Preston, Julia. "The Truth About Mexican-Americans." *New York Review of Books*, December 3, 2015. http://www.nybooks.com/articles/2015/12/03/truth-about-mexican-americans/.

Rincon, Paul. "Ancient Britons 'replaced' by newcomers." BBC *News*, February 21, 2018. https://www.bbc.com/news/science-environment-43115485.

Roberts, Kenneth. *Northwest Passage*. Garden City, NY: Doubleday, Doran and Company, 1937.

Romero, Simon. "Pope's Trip to Brazil Seen as 'Strong Start' in Revitalizing Church." *New York Times*, July 28, 2013. http://www.nytimes.com/2013/07/29/world/americas/vibrant-display-at-popes-last-mass-in-brazil.html.

Ruff, Abdul. "Genocides of Tamils and Indo-Sri Lanka relations." *Modern Diplomacy*, April 3, 2017. https://moderndiplomacy.eu/2017/04/03/genocides-of-tamils-and-indo-sri-lanka-relations/.

Salisbury, David S., A. William Flores De Melo, and Pedro Tipula Tipula. "Transboundary Political Ecology in the Peru-Brazil Borderlands: Mapping Workshops, Geographic Information, and Socio-Environmental Impacts." *Revista Geográfica* 152 (2012): 105–15. http://scholarship.richmond.edu/cgi/viewcontent.cgi?article=1032&context=geography-faculty-publications.

Salomon, Richard. "On the Origin of the Early Indian Scripts: A Review Article." *Journal of the American Oriental Society* 115, no. 2 (1995): 271–9.

Schiavenza, Matt. " 'Well Done': The Legalization of Gay Marriage in Ireland." *Atlantic*, May 23, 2015. https://www.theatlantic.com/international/archive/2015/05/ireland-gay-marriage/394052/.

Seelow, Soren. "Ce massacre a été occulté de la mémoire collective: L'exécution de plus d'une centaine de manifestants algériens à Paris le 17 octobre 1961 reste méconnue, souligne l'historien Gilles Manceron." *Le Monde*, October 17, 2011. http://www.lemonde.fr/societe/article/2011/10/17/17-octobre-1961-ce-massacre-a-ete-occulte-de-la-memoire-collective_1586418_3224.html.

Sellstrom, Tor, and Lennart Wohlgemuth. *The International Response to Conflict and Genocide: Lessons from the Rwanda Experience* (JEEAR). Uppsala, SE: The Nordic Africa Institute, 1996.

Shah, Inayat. "Linguistic Attitude and the Failure of Irish Language Revival Efforts." *International Journal of Innovation and Scientific Research* 1, no. 2 (2014). http://www.ijisr.issr-journals.org/abstract.php?article=IJISR-14-111-02.

Shivaram, Choodie. "Where Women Wore the Crown: Kerala's Dissolving Matriarchies Leave a Rich Legacy of Compassionate Family Culture." *Hinduism Today*, February 1996. https://www.hinduismtoday.com/modules/smartsection/item.php?itemid=3569.

Skinner, Brett J. "Medicare, the Medical Brain Drain and Human Resource Shortages in Health Care." AIMS *Health Care Reform Background Paper #7* (December 2002). http://www.aims.ca/site/media/aims/BrainDrain.pdf.

Smith, Danielle. "Alberta Already Was an NDP Province." *Globe and Mail*, May 8, 2015. http://www.theglobeandmail.com/opinion/alberta-already-was-an-ndp-province/article24316040/.

Soderstrom, Mary. *Green City: People, Nature and Urban Places*. Montreal: Véhicule Press, 2006.

———. *Making Waves: The Continuing Portuguese Adventure*. Montreal: Véhicule Press, 2010.

———. "Old Time Fiddlers Make Beautiful Music in Vermont." *New York Times*, May 6, 1973.

———. *Road Through Time: The Story of Humanity on the Move*. Regina: University of Regina Press, 2017.

———. *The Violets of Usambara*. Toronto: Cormorant Books, 2008.

Somerset, A.J. "How Canadians Helped Create the NRA." *Toronto Star*, December 20, 2015. Excerpt from *Arms: The Culture and Credo of the Gun* by A.J. Somerset (Windsor, ON: Biblioasis, 2015). https://www.thestar.com/news/insight/2015/12/20/how-canadians-helped-create-the-nra.html.

Speke, John Hanning. *Journal of the Discovery of the Source of the Nile*. London: W. Blackwood and Sons, 1863; Project Gutenberg, 2002. http://www.gutenberg.org/files/3284/3284.txt.

Stewart, Walter. *The Life and Political Times of Tommy Douglas*. Toronto: McArthur and Company, 2003.

Strabo. *Geography. Volume 1: Books 1–2.* Translated by Horace Leonard Jones. Loeb Classical Library. Cambridge, MA: Harvard University Press, 1917. Available online at http://penelope.uchicago.edu/Thayer/e/roman/texts/strabo/2e1*.html.

Springhall, John. " 'Kicking out the Vietminh': How Britain Allowed France to Reoccupy South Indochina, 1945–46." *Journal of Contemporary History* 40, no. 1 (2005): 115–30.

Stegner, Wallace. *Wolf Willow: A History, a Story and a Memory of the Last Plains Frontier.* New York: Viking Press, 1962.

Summers, Hannah. "Yemen on brink of 'world's worst famine in 100 years' if war continues." *Guardian,* October 15, 2018. https://www.theguardian.com/global-development/2018/oct/15/yemen-on-brink-worst-famine-100-years-un.

Taylor, K.W. *A History of the Vietnamese.* Cambridge: Cambridge University Press, 2013.

Theroux, Paul. *Ghost Train to the Eastern Star: On the Tracks of the Great Railway Bazaar.* Boston and New York: Houghton Mifflin, 2008.

Thompson, Derek. "Canada's Secret to Escaping the 'Liberal Doom Loop.' " *Atlantic,* July 9, 2018. https://www.theatlantic.com/ideas/archive/2018/07/canadas-secret-to-escaping-the-liberal-doom-loop/564551/.

Thornton, John. *Africa and Africans in the Making of the Atlantic World, 1400–1800.* Cambridge: Cambridge University Press, 1998.

Tishkoff, Sarah A., Floyd A. Reed, Alessia Ranciaro, Benjamin F. Voight, Courtney C. Babbitt, Jesse A. Silverman, Kweli Powell, Holly M. Mortensen, Jibril B. Hirbo, Maha Osman, Muntaser Ibrahim, Sabah A. Omar, Godfrey Lema, Thomas B. Nyambo, Jilur Ghori, Suzannah Bumpstead, Jonathan K. Pritchard, Gregory A. Wray, and Panos Deloukas. "Convergent adaptation of human lactase persistence in Africa and Europe." *Nature Genetics* 39, no. 1 (2007). https://doi.org/10.1038/ng1946.

Toronto Star. "Stephen Harper's most controversial quotes compiled—by Tories." *Toronto Star,* April 25, 2011. https://www.thestar.com/news/canada/2011/04/25/stephen_harpers_most_controversial_quotes_compiled_by_tories.html.

Trombetta, Beniamino, Eugenia D'Atanasio, Andrea Massaia, Marco Ippoliti, Alfredo Coppa, Francesca Candilio, Valentina Coia, Gianluca Russo, Jean-Michel Dugoujon, Pedro Moral, Nejat Akar, Daniele Sellitto, Guido Valesini, Andrea Novelletto, Rosaria Scozzari, and Fulvio Cruciani. "Phylogeographic refinement and large scale genotyping of human Y chromosome haplogroup E provide new insights into the dispersal of early pastoralists in the African continent." *Genome Biology and Evolution* 7, no. 7 (2015). https://doi.org/10.1093/gbe/evv118.

Turits, Richard Lee. "A World Destroyed, A Nation Imposed: The 1937 Haitian Massacre in the Dominican Republic." *Hispanic American Historical Review* 82, no. 3 (2002): 589–635.

Valdman, Albert. "Creole: The National Language of Haiti." *Footsteps* 2, no. 4 (2000). http://www.indiana.edu/~creole/creolenatllangofhaiti.html.

Vargas Llosa, Mario. *The Feast of the Goat.* Translated by Edith Grossman. New York: Picador, 2005.

Viotti da Costa, Emilia. *The Brazilian Empire: Myths and Histories*. Chapel Hill and London: University of North Carolina Press, 2000.

Wade, Lizzie. "South Asians are descended from a mix of farmers, herders, and hunter-gatherers, ancient DNA reveals." *Science Magazine*, April 18, 2018. https://www.sciencemag.org/news/2018/04/south-asians-are-descended-mix-farmers-herders-and-hunter-gatherers-ancient-dna-reveals.

Waiser, Bill. *A World We Have Lost: Saskatchewan Before 1905*. Markham, ON: Fifth House, 2016.

Wheeler, James Talboys. *India Under British Rule from the Foundation of the East India Company*. London: Macmillan, 1886; Project Gutenberg, 2014. http://www.gutenberg.org/files/46151/46151-h/46151-h.htm.

Whittington, Les. "2014 Canadian budget expected to keep 'starving the beast.'" *Toronto Star*, January 17, 2014. https://www.thestar.com/news/canada/2014/01/17/2014_canadian_budget_expected_to_keep_starving_the_beast.html.

Willis, Michael J. *Politics and Power in the Maghreb: Algeria, Tunisia and Morocco from Independence to the Arab Spring*. New York, Columbia University Press, 2012.

Woodard, Roger D. "*Phoinikēia Grammata*: An Alphabet for the Greek Language." In *A Companion to the Ancient Greek Language*, edited by Egbert J. Bakker, 25–46. Malden, MA: Wiley-Blackwell, 2010.

Wooster, Chuck. "Glaciers and Taxes in Vermont and New Hampshire." *Northern Woodlands*, May 12, 2002. https://northernwoodlands.org/outside_story/article/glaciers-and-taxes-in-vermont-and-new-hampshire.

———. "Vermont—New Hampshire: The Hydropower Difference." *Northern Woodlands*, April 14, 2002. https://northernwoodlands.org/outside_story/article/vermont-new-hampshire-the-hydropower-difference.

Wu, Huizhong. "India's Cheraman mosque: A symbol of religious harmony." *Al Jazeera*, April 23, 2017. https://www.aljazeera.com/indepth/features/2017/04/india-cheraman-mosque-symbol-religious-harmony-170406095923455.html.

Wucker, Michele. "The Dominican Republic's Shameful Deportation Legacy." *Foreign Policy*, October 8, 2015. http://foreignpolicy.com/2015/10/08/dominican-republic-haiti-trujillo-immigration-deportation/.

Younglai, Rachelle. "Canadian women more active in labour force than American peers." *Globe and Mail*, August 17, 2016. http://www.theglobeandmail.com/report-on-business/economy/canadian-womens-work-force-participation-surges/article31450579/.

Zovatto, Daniel. "Dominican Republic opts for continuity." *Brookings*, June 2, 2016. https://www.brookings.edu/opinions/dominican-republic-opts-for-continuity-2/.

INDEX

Mary Soderstrom is the author of numerous books, including *Road Through Time: The Story of Humanity on the Move* and *Desire Lines: Stories of Love and Geography*. She lives in Montreal.